Historical Geography and Geosciences

This book series serves as a broad platform for contributions in the field of Historical Geography and related Geoscience areas. The series welcomes proposals on the history and dynamics of place and space and their influence on past, present and future geographies including historical GIS, cartography and mapping, climatology, climate history, meteorology and atmospheric sciences, environmental geography, hydrology, geology, oceanography, water management, instrumentation, geographical traditions, historical geography of urban areas, settlements and landscapes, historical regional studies, history of geography and historic geographers and geoscientists among other topically related areas and other interdisciplinary approaches. Contributions on past (extreme) weather events or natural disasters including regional and global reanalysis studies also fit into the series.

Publishing a broad portfolio of peer-reviewed scientific books Historical Geography and Geosciences contains research monographs, edited volumes, advanced and undergraduate level textbooks, as well as conference proceedings. This series appeals to scientists, practitioners and students in the fields of geography and history as well as related disciplines, with exceptional titles that are attractive to a popular science audience.

If you are interested in contributing to this book series, please contact the Publisher.

More information about this series at http://www.springer.com/series/15611

Bruno Schelhaas • Federico Ferretti •
André Reyes Novaes •
Marcella Schmidt di Friedberg
Editors

Decolonising and Internationalising Geography

Essays in the History of Contested Science

IGU UGI

Springer

Editors
Bruno Schelhaas
Archive for Geography
Leibniz Institute for Regional Geography
Leipzig, Germany

Federico Ferretti
School of Geography
University College Dublin
Belfield, Dublin 4, Ireland

André Reyes Novaes
Department of Human Geography
Rio de Janeiro State University
Rio de Janeiro, Brazil

Marcella Schmidt di Friedberg
"Riccardo Massa" Department
of Educational Human Sciences
University of Milano-Bicocca
Milan, Italy

ISSN 2520-1379 ISSN 2520-1387 (electronic)
Historical Geography and Geosciences
ISBN 978-3-030-49515-2 ISBN 978-3-030-49516-9 (eBook)
https://doi.org/10.1007/978-3-030-49516-9

This Springer imprint is published by the registered company Springer Nature Switzerland AG
The registered company address is: Gewerbestrasse 11, 6330 Cham, Switzerland

Introduction: Decolonising and Internationalising Geography—Trends and Challenges Towards Individuality and Plurality

This collective book is the result of two symposia organised by the Commission History of Geography of the International Geographical Union in July 2017. These sessions took place at the occasion of the 17th International Congress of History of Science and Technology, in the suggestive location of the Praia Vermelha Campus of the Federal University of Rio de Janeiro (UFRJ), with, respectively, the titles of "Geography as an international science" and "Critical, radical and postcolonial geographies and cartographies from early approaches to present-day debates". These sessions were attended by scholars from Brazil, France, Germany, Russia, USA, Japan, Argentina, Chile, Mexico, Spain, Italy, Switzerland, Ireland and UK, and the successive debates highlighted the existence of different ideas on two big and polysemic words such as internationalism and de(post)colonialism. This introduction does not aim to define these words or explore their genealogies in geographical knowledge, but just to address some general trends and challenges that scholars involved in these circuits, especially historians of geography, are facing when they articulate their investigations with these terms.

Among the attendants of these symposia, those who gave their availability to participate in this book were invited by the editors to transform their presentations in chapters, consistently with the politics deployed by the Commission History Geography for fostering international scholarship and increasing the space for the history and philosophy of geography in the academic world and in international literature. The resulting book is divided in two parts, corresponding to two broad problematics that proved anyway to be widely discussed in scholarly debates in the field of historical and critical geographies in the last years, especially under the form of militant wishes. Indeed, for most of us, the problem is not merely to analyse internationality or decoloniality in geography: what we want is to internationalise and decolonise our discipline, with all the possible challenges and contradictions annexed, as follows.

From Decolonialism to Internationalism, and Return

We first consider that trends and challenges that made geographical knowledge move on are always situated in space and time. If it is well known that knowledge has a mobility, then it is important to consider the meanings of words in different disciplines, places and periods. Geography was definitively not a pioneer field to discussing the notions of internationalism and de(post)colonialism, which were first raised in other disciplines. However, these two words are now a central concern among many geographers, mainly among those interested in historical geography and the history of geography. This also resulted in recent debates and publications that may define important landmarks to understand how the term has been used and conceived in a disciplinary context. In the same year of the international conference in Rio de Janeiro, conferences and projects developed in the context of British geography can also illustrate the growing importance of internationalist/transnational and decolonial approaches in geography. It was especially the case with the 2017 Annual International Conference of the Royal Geographical Society with Institute of British Geographers, which give a further example of the disputed meanings that surrounds the choose of key terms and topics in scientific knowledge. The theme of that conference was eventually "Decolonising geographical knowledge: opening geography out to the world" (Radcliffe 2017).

Concurrently, the research project on historical geographies of internationalism led by authors based at the School of Geography at Nottingham University (Stephen Legg, Mike Heffernan, Jake Hodder and Benjamin Thorpe) produced a series of meetings, exhibitions and publications on international journals to study geography's internationalism. In their introduction to the special issue on this matter that they edited for *Political Geography*, these authors argued that, "although geographers have only made a modest contribution" to study internationalism, "they have a significant role to play" (Hodder et al. 2015: 1). Drawing on the work of the professor of International Relations Halliday (1988), the authors seek to understand how different kinds of internationalism have been devised, deployed and enacted. The main aim of their project is to explore several ways in which a geographical perspective can contribute to "rethinking the international", asking "who articulates the international and from where".

These works parallel an established French–German literature which has addressed the history of the international geographical congresses explicitly calling for the adoption of pluralistic methodological approaches and conceptual tools, that might be those of global history, the transnational turn, the *histoire croisée* or *histoire comparée* (Robic 2013; Debarre 2014), in analysing transfers of knowledge between different linguistic areas. While scholarship on international conferences has shown that internalisation is not automatic in intellectual life and can be challenging, several cases show anyway that intellectual exchanges between scholars of two or more states can continue even in the context of geopolitical tensions between their respective nations, as it occurred in continental Europe between the nineteenth and the twentieth century. This allows the authors use definitions such as "the international fabrication of geography" (Debarre 2014: 12). If places do something to knowledge (Livingstone 2003), then also transfers, transits and circulations transform knowledge in its sifting through places and cultures (Secord 2004). Therefore, internationalism and transnationalism should be able to make sense of local histories and specificities as well as of complexity and pluralism.

In this vein, such approaches to internationalism can be easily connected with decolonial scholarship, which placed the geopolitics of knowledge's production in the centre of the critical debates about colonialism, eventually the Latin American Modernity–Coloniality–Decoloniality (MCD) collective (Quijano 2000; Mignolo 2002). The MCD group's main argument is that modernity co-emerged with coloniality, which means that colonial power is never exceptional and situated "elsewhere", but should be understood considering socio-spatial relations across multiple scales. MCD authors are akin to strands of research on post-development, with which they share a critique of the adoption of Northern recipes and notions for the solution of the problems of the South. Authors such as Walter Mignolo and Arturo Escobar especially criticise what they deem Eurocentric critiques of Eurocentrism, including poststructuralist inflected versions of postcolonial studies, overwhelmingly drawing upon European authors and concepts (Mignolo and Escobar 2010). Therefore, MCD authors call for the reconsideration of sources from the Global South (or Souths), including indigenous authors and activists. Importantly, this does not result in a provincialism arguing for completely delinking with Northern traditions, but rather in a pluralistic view where, rather than cultural "purity", one values notions such as "border thinking" and "trans-modernity" to engage in a dialogue beyond the essentialisation of cultures. Significantly, one of the latest notions on which these authors work is that of "pluriverse", which is defined as, after the Zapatistas, "a world where many worlds fit" (Kothari et al. 2019: xxviii) proposing a very pluralistic series of possible pathways for critical thinking and pluralism in the Souths as well as in the Norths.

Thus, rather than producing a Eurocentric critique of Eurocentrism, decolonial approaches seek for a critique of Eurocentrism from subalternised and silenced knowledge. As Grosfoguel (2011) argued, this is not a fundamentalist, anti-European critique but an attempt to go beyond both hegemonic and marginal fundamentalisms to reframe critical thinking about the universal and the particular. This interest of MCD authors in scales, global–local co-constitution and travelling knowledge could clearly speak to the concerns of decolonial scholarship. If the debates on internationalism seek to propose a geographical approach that "examine spaces and sites not previously considered in international histories" (Hodder et al. 2015: 1), then a solid bridge could be constructed between approaches on internationalism and decolonialism, including recent studies on geography and postcolonialism arguing for the importance of the "planetary" and for the consideration of "other" postcolonialisms, including the importance of translation (Sidaway et al. 2014).

It is not a coincidence that some authors involved in the discussion of internationalism in British geography were also engaged in the debates for "decolonising geography" (Jazeel 2017; Legg 2017), stimulated by the Chair's theme of the 2017 RGS-IBG international conference mentioned above. Comments and reflections about this key problematic were published in relevant journals, such as *Area* and *Transactions of the Institute of British*

Geographers, demonstrating how this conference was important in the circulation of the term "decolonial" among this group of geographers. The Chair of the Conference, Sarah Radcliffe, seeks to examine the ways in which decolonial scholarship seeks to build on—and go beyond postcolonialism to provide an "overview of decolonising approaches for [British] geographers unfamiliar with the field" (Radcliffe 2017: 1). Postcolonial geography has identified the longstanding legacies of colonialism and critiqued metropolitan schemas by exploring the ways of imagining and describing the world. According to Radcliffe, decolonial approaches emerged from and engaged with a wide range of critical and radical scholarship, such as Black theory, Indigenous theory, feminist and queer theory, including the MCD group. This last group of authors, mainly Latin Americans working in different parts of the world, is increasingly influential in the growing circulation of the term decolonialism, in recent decades, among English-speaking geographers.

As Radcliffe (2017) recognises, geography has not engaged with these topics for much of its disciplinary history. Moreover, the circulation of decolonial matters and terminologies among British geographers was not free of criticism. Most of the 2017 plenaries at the RGS-IBG, for instance, were harshly criticised by geographers who claimed that decolonial approaches could not be restricted to epistemological or theoretical approaches in knowledge production. In their commentary about the conference published for *Area*, Esson et al. (2017) argue that a merely theoretical critical consciousness acquired via decolonial thinking could do more harm than good, because it could end with reproducing coloniality. The emphasis in decolonising geographical knowledge rather than structures, practices and institutions and the idea to "open geography on to the world" without major questionings of geography's institutions seems very problematic for these authors. They namely argue for the real inclusion of categories such as non-White and non-academic people which were traditionally excluded by the places of production of scholarly geography. If decolonisation is not a "metaphor" (Tuck and Yang 2012) or a theme that can "be taken up for three days of a conference and then put down" (Esson et al. 2017: 385), then its circulation presents a series of challenges for geographical practices in academia.

The colonialist structures inherited from geography's colonial and imperial pasts have many durabilities (Stoler 2016) and, as Esson et al. (2017: 387) note, the discipline "may not be ready to, or even capable of, responding to the challenge of decolonisation". In fact, the majority of geographers—both in Britain and in South America – remain external to contemporary decolonial struggles and there are many challenges to achieve what Rivera Cusicanqui (2012) called "decolonising practices", which should go deeper than the circulation of a new vocabulary for critical studies on colonialism. The risk here is to reproduce colonialism and recentring non-indigenous, White and otherwise privileged groups in the global architecture of knowledge production (Esson et al. 2017). In fact, facing the challenge of recognising the agency of those who have been racialised as indigenous and non-White by coloniality is not easy. As Esson et al. (2017:

386) highlighted, by engaging with decolonial approaches, geographers "run the risk of speaking not for but instead" of those who were often excluded of the international histories of the discipline.

International and Multilingual Networking for Disciplinary Pluralism

All these topics and challenges underpin the debates among the scholars who presented their investigations at the 25th International Congress of the History of Science and Technology held in Rio de Janeiro in July 2017. The chapters of this book will not discuss directly these challenges, because they offer empirical investigations made from different parts of the world with multiple interests. Nevertheless, the empirical proposition offered is all connected with different approaches on internationalism and decolonialism. They may work as an empirical floor to discuss how decolonial scholarship could be presented in the research practices concerning different subjects and topics such as language, planning, mapping and teaching.

It is worth noting that these two symposia were not an occasional event, because they happened in the context of an established international networks interested in doing critical histories of geographical knowledge, one which gathers periodically around the meetings organised by the IGU Commission History of Geography. Locally, this commission permanently collaborates with the Brazilian Network on the History of Geography and on Historical Geography. These history of geography meetings are almost a unique case in the panorama of international conferences, as an effective multilingualism is put in place there, given that most of attendants are familiar with several languages (typically English plus one or more Latin languages). As a result, most of the presentations were given in idioms such as Portuguese, Spanish and French, with the occasional use of English as lingua franca, so that there was great scope for mutual understanding without cancelling linguistic difference. While this is not unproblematic and could not necessarily be replicated in all situations, it is an important demonstration that international scholarly cooperation can challenge all forms of monolingualism and monoculturalism (from dominant English to other languages representing regional imperialisms), which constitute a serious limitation to all the endeavours for effectively decolonising geographical knowledge and practices.

In the first part of the book "Decolonising and Internationalising Geography", André Reyes Novaes reassesses in postcolonial ways the views of Brazilian indigenous by Portuguese historian Jaime Cortesão. Maria Verónica Ibarra and Edgar Telledos Sanchez do a comparative analysis of the main works of two "pioneers" of critical geographies in Latin America, the Brazilian Josué de Castro and the Cuban António Nuñez Jiménez. Akio Onjo analyses the indirect effects of war in shaping Japanese national identities through the politics for sick and wounded soldiers. Marcella Schmidt di Friedberg and Stefano Malatesta analyse recent geopolitical issues involving the Maldives and Chagos Islands in the context of postcolonial Indian Ocean. Pascal Clerc questions the North/South divide based on its representation in French public discourses and didactic materials. Finally, Verónica Hollman analyses the ways in which drone photography is reshaping imaginations of nature. In the second part, Maximilian Georg and Ute Wardenga present a research project ongoing at the *Leibniz Institute for Regional Geography* at Leipzig interrogating notions of geography's internationalism through international comparison of geographical societies established between 1821 and 1914. Toshiuky Shimazu addresses geography's internationalism in the cultural context of late nineteenth-century Paris. Larissa Alves de Lira analyses the work of French geographer Pierre Monbeig in Brazil through the lenses of "geo-history". Mariana Lamego questions commonplaces on disciplinary internationalisation around the 1956 International Geographical Congress celebrated in Rio de Janeiro. Bruno Schelhaas and Stephan Pietsch finally analyse the problems of writing the history of the International Geographical Union starting from its archives. We present this heterogeneous set of texts as a contribution to internationalise and decolonise geography, pluralistically and pluriversally.

Bruno Schelhaas
Federico Ferretti
André Reyes Novaes
Marcella Schmidt di Friedberg

References

Débarre S (ed) (2014) Géographies entre France et Allemagne – Acteurs, notions et pratiques (fin XIXe siècle – milieu XXe siècle). Revue Germanique Internationale 20 (dossier thématique)

Esson J, Noxolo P, Baxter R, Daley P, Byron M (2017) The 2017 RGS-IBG chair's theme: decolonising geographical knowledges, or reproducing coloniality? Area 49(3):384–388

Grosfoguel R (2011) Decolonizing post-colonial studies and paradigms of political-economy: transmodernity, decolonial thinking, and global coloniality. Transmodernity: J Peripheral Cult Prod Luso-Hispanic World 1(1):1–38

Halliday F (1988) Three concepts of internationalism. Int Aff 64:187–198

Hodder J, Legg S, Heffernan M (2015) Introduction: historical geographies of internationalism, 1900–1950. Polit Geogr 49:1–6

Jazeel T (2017) Mainstreaming geography's decolonial imperative. Trans Inst Brit Geogr 42(3):334–337

Kothari A, Salleh A, Escobar A, Demaria F, Acosta A (2019) Pluriverse, a post-development dictionary. Tulika Books and Authorsupfront, New Delhi

Legg S (2017) Decolonialism. Trans Inst Brit Geogr 42(3):345–348

Livingstone DN (2003) Putting science in its place. Chicago University Press, Chicago

Mignolo W (2002) The geopolitics of knowledge and the colonial difference. The S Atlantic Q 101(1):57–96

Mignolo W, Escobar A (eds) (2010) Globalization and the decolonial option. Routledge, London

Quijano A (2000) Coloniality of power and Eurocentrism in Latin America. Int Sociol 15 (2):215–232

Radcliffe S (2017) Decolonising geographical knowledges. Trans Inst Brit Geogr 42:329–333

Rivera Cusicanqui S (2012) Ch'ixinakax utxiwa: a reflection on the practices and discourses of decolonization. S Atlantic Q 111:95–109

Robic MC (2013) À propos de transferts culturels. Les congrès internationaux de géographie et leurs spatialités. Revue germanique internationale 12:33–45

Secord J (2004): Knowledge in transit. Isis 95(4):654–672

Stoler AL (2016) Duress: Imperial durabilities for our times. Duke University Press, London

Sidaway J, Woon CY, Jacobs J (2014) Planetary postcolonialism. Singap J Trop Geogr 35:4–21

Tuck E, Yang KW (2012) Decolonization is not a metaphor. Decolonization Indigeneity Educ Soc 1:1–40

Contents

About the Editors

Bruno Schelhaas received his Ph.D. from the University of Leipzig. He is the Head of the Archive for Geography at Leibniz Institute for Regional Geography in Leipzig and is the archivist of the International Geographical Union. His interests include the history of geography and cartography, historical geography and archival science. He is member of the Steering Committee of the Commission History of Geography of the International Geographical Union.

Federico Ferretti received his Ph.D. from the Universities of Bologna and Paris 1 Panthéon-Sorbonne in 2011. After research and teaching experiences in Italy, France, Switzerland and Brazil, he is serving as an Associate Professor at UCD School of geography, working in the fields of philosophy and history of geography and on critical and anarchist geographies with a special focus on Latin America. He authored, co-authored or edited 15 books in Italian, French and English and published research papers in the major international peer-reviewed journals in his area of study, in English, French, Italian, Spanish and Portuguese. He is the Secretary/Treasurer for the Commission History of Geography of the International Geographical Union, Secretary of the History and Philosophy of Geography Research Group of the RGS-IBG and member of the Geosciences and Geographical Sciences Committee of the Royal Irish Academy.

André Reyes Novaes is an Associate Professor at the Rio de Janeiro State University. He is currently Honorary Research Associate in the Department of Geography at the Royal Holloway University of London and member of the Steering Committee of the Commission History of Geography of the International Geographical Union. He is on the editorial team of the book series *Geographers: Biobibliographical Studies* and the journal *Espaço e Cultura*, a pioneering publication on cultural approaches in human geography in Brazil. His research interests include visual methods, history of cartography, popular geopolitics and history of South American borders.

Marcella Schmidt di Friedberg is full Professor of Geography, at the "Riccardo Massa" Department of Human Sciences for Education (University of Milano-Bicocca, Italy). She is Chair from 2016 of the International Geographical Union Commission on History of Geography and Vice-Director of the Marine Research and High Education Center in

Faaf-Magoodhoo (Rep. of Maldives). Her research interests concern Cultural Geography, Hazard and Resilience, Gender Geography and History of the Geographical Thought. She has been working especially on the relations between nature, culture, memory and landscape in different contexts, from the Mediterranean to Japan.

Mapping Cross-Cultural Exchange: Jaime Cortesão's Dialogues and Documents on the Role of Indigenous Knowledge in Brazilian Exploration

1

André Reyes Novaes

Abstract

This chapter aims to explore Jaime Cortesão's textual and visual narratives about indigenous knowledge. Although many scholars investigated Cortesão's production during his exile in Brazil (1940–1957), little attention has been given to study how he investigated the role of indigenous knowledge in territorial exploration and mapping. To address this topic, I will identify debates and documents explored by Cortesão to approach indigenous maps. The chapter is divided into two sections. First, I will explore Cortesão's dialogues, considering how academic debates and references encouraged him to study indigenous people's spatial knowledge. Then, I will stress some textual and visual documents selected by Cortesão to discuss indigenous maps as a specific category. In addition to explorer's narratives, Cortesão presented a group of maps from 1721 to 1724, discovered by him at the National Library of Rio de Janeiro and classified as *bandeirantes* due to its indigenous influence. By identifying Cortesão's dialogues and documents, my primary intention here is to discuss his comments on indigenous mapping in the light of postcolonial and decolonial approaches to understanding exploration maps as co-produced and hybrid artefacts.

Résumé

Ce chapitre vise à explorer les discours textuels et visuels de Jaime Cortesão sur les savoirs autochtones. Bien que de nombreux chercheurs aient étudié la production de Cortesão au cours de son exil au Brésil (1940–1957), ses travaux sur le rôle des connaissances autochtones dans l'exploration et la cartographie ont été moins pris en compte. Afin d'aborder ce sujet, je vais identifier les débats et les documents explorés par Cortesão pour traiter des cartes indigènes. Le chapitre est divisé en deux parties. Dans un premier temps, je vais explorer les dialogues de Cortesão, en examinant comment les débats universitaires et les références l'encouragent à étudier les connaissances spatiales des peuples autochtones. Ensuite, je soulignerai certains documents textuels et visuels sélectionnés par Cortesão pour discuter des cartes indigènes en tant que catégorie spécifique. En plus des récits de l'explorateur, Cortesão a présenté un groupe de cartes de 1721 et 1724, fondé par lui à la Bibliothèque

A. R. Novaes (✉)
Departamento de Geografia Humana, Universidade do Estado do Rio de Janeiro, Rio de Janeiro, Brazil
e-mail: andrereyesnovaes@gmail.com

B. Schelhaas et al. (eds.), *Decolonising and Internationalising Geography*, Historical Geography and Geosciences, https://doi.org/10.1007/978-3-030-49516-9_1

nationale de Rio de Janeiro et classé comme *bandeirantes* en raison de son influence indigène. En identifiant les dialogues et les documents de Cortesão, mon intention principale ici est de discuter de ses commentaires sur la cartographie indigène à la lumière des approches postcoloniales et décoloniales pour comprendre les cartes d'exploration en tant qu'artefacts coproduits et hybrides.

Resumen

Este capítulo tiene como objetivo explorar las narraciones textuales y visuales de Jaime Cortesão sobre el conocimiento indígena. Aunque muchos académicos investigaron la producción de Cortesão durante su exilio en Brasil (1940–1957), se ha prestado menos atención a sus estudios sobre el papel del conocimiento indígena en la exploración y mapeo territorial. Para abordar este tema, identificaré debates y documentos explorados por Cortesão para abordar mapas indígenas. El capítulo está dividido en dos secciones. Primero, exploraré los diálogos de Cortesão, considerando cómo los debates académicos y las referencias lo alentaron a estudiar el conocimiento espacial de los pueblos indígenas. Luego, destacaré algunos documentos textuales y visuales seleccionados por Cortesão para discutir los mapas indígenas como una categoría específica. Además de las narraciones de los exploradores, Cortesão presentó un grupo de mapas de 1721 y 1724, encontrados por él en la Biblioteca Nacional de Río de Janeiro y clasificado como *bandeirantes* debido a su influencia indígena. Al identificar los diálogos y documentos de Cortesão, mi intención principal es discutir sus comentarios sobre el mapeo indígena a la luz de los enfoques poscoloniales y decoloniales para entender los mapas de exploración como artefactos coproducidos e híbridos.

Keywords

Indigenous knowledge · Jaime Cortesão · Exploration maps · Hybrid artefacts · *Bandeirantes*

1.1 Introduction

In his book called the History of Brazil in Old Maps [*História do Brasil nos Velhos Mapas*], the Portuguese scholar Jaime Cortesão (1884–1960) gathered a series of documents to support his comments on the exploration of Brazilian territory. In the fourth part of the second volume, the author listed archives related to the cartography of Indians and explorers, considering a series of maps produced in the eighteenth century to portray "territories that had just been occupied and incorporated to Brazilian history" (Cortesão 2009: 231).

Cortesão made specific reference to a group of maps commented in a letter written by the Jesuit Priest Diogo Soares to the King of Portugal Dom João V. In June 1730, the priest was in charge to draw a "new atlas of Brazil" and to accomplish the task he referred to some maps recently made by explorers. Cortesão searches for these maps at Evora's library in Portugal. However, it was during his exile in Brazil between 1940 and 1957 that he found a group of maps at the National Library of Rio de Janeiro. The cartographic materials were rustic, limited to the representation of rivers, drawn in rudimentary papers, and dated from 1721 to 1724, what lead Cortesão to conclude that he had found the maps mentioned by the Jesuit priest.

Cortesão found the same maps, but proposed a new issue: who made these maps? Diogo Soares called the mapmakers *sertanistas*, considering those who systematically explored the *sertão* (countryside) with scientific proposes.

Cortesão defended the authorship for the *bandeirantes*, a term applied to identify explorers who depart from São Paulo seeking for gold mines.[1] Considering this controversy about maps authorship, Cortesão (2009: 232) recognized the "need for a specific method" to differentiate these documents. His method had a clear hypothesis: there is a more substantial indigenous influence in the cartography produced by *bandeirantes*, which shows how a cross-cultural exchange of geographic knowledge was essential for the success of exploration.

Cortesão's narrative met the demand for a conciliatory history about Brazilian colonization and highlighted São Paulo's role in shaping national identity (Raimundo 2004; Ribeiro 2018). However, his investigation has also offered inspiring approaches to study the role of indigenous knowledge in territorial exploration, subject of interest to many contemporary scholars. Indeed, the topic explored by Cortesão could sound quite up-to-date nowadays, because the agency of indigenous peoples in guiding expeditions has been a central concern in contemporary historical geography and history of cartography. By seeking to challenge dominant narratives in the history of exploration, which privileges the actions of heroic individuals, historical geographers have drawn attention to the cross-cultural exchange of geographical knowledge (Driver and Jones 2009). To rewrite histories of expeditions focusing on the encounters and interactions, these scholars proposed to understand exploration as knowledge transfer (Lefebvre and Surun 2008) and turn our attention to "indigenous intermediaries" (Konishi et al. 2015).

In the same direction, historians of cartography also highlighted the agency of indigenous knowledge in territorial mapping to challenge hegemonic narratives in their discipline. To criticize conceptions of cartography as a "marvel of European science" (Brückner 2011: 4) or a monolithic endeavour (Edney 1993), historians of cartography also emphasized the role of cross-cultural exchanges in map production, circulation, and consumption. To replace persistent myths on discovery and mapmaking, scholars have been exploring the notion of "cartographic encounters" (Short 2009), considering how the successful European exploration of the New World resulted from the exchange of information between newcomers and indigenous people.

Inspired by postcolonial and decolonial approaches that have become prominent in the social sciences since the 1990s, scholars from different backgrounds revisit and critically reinterpret archives searching for an indigenous agency (Cusicanqui 2015). However, as Cortesão's writings could show, the attention to indigenous knowledge is not an exclusivity of contemporary approaches. In different circumstances, indigenous knowledge was classified, studied, and appropriated in historical narratives. As Driver (2016) pointed out, the terms varied over time, and the "native information" of the nineteenth century becomes the "indigenous knowledge" of the late twentieth century. Considering how these categories were defined in different historical contexts, this chapter aims to explore how indigenous knowledge was presented in Cortesão's textual and visual narratives. Although many scholars already explored Cortesão's abundant production on "Brazilian topics" during his exile (Oliveira 2014a; Oliveira 2015), few comments have been made about his approaches to investigate the role of indigenous knowledge in territorial exploration (Ribeiro 2018). To explore this topic, I will identify debates and documents explored by Cortesão to deal with indigenous maps and exploration practices.

Drawing on selected writings and images, I would like to propose three main questions: which theoretical references and debates

[1] The difference between *bandeirantes* and *sertanistas* is understood in this chapter as a symbolic construction that influenced many historical narratives about the exploration of Brazilian territory. Jaime Cortesão stated that *bandeirantes* were "sertanistas from São Paulo", "shaped by the lifestyle of the *bandeirismo*" (Cortesão 2009: 231). The *sertanistas* were "Luso-Brazilians from other captaincies", official servants who map the territory attending state demands. This differentiation was crucial in Cortesão's argument and, as I will argue in this chapter, it must be understood through the author's relationship with São Paulo's elites and their use of *bandeirismo* as a local identity symbol.

stimulated Cortesão to discuss the indigenous role in territorial exploration? What kinds of historical documents did the author select to explore this topic? What are the potentialities and limitations offered by Cortesão's methodologies for contemporary decolonial and postcolonial approaches engaging with "native" or "indigenous" maps?

Aware of the risks of anachronism, this chapter seeks to build bridges between postcolonial and decolonial approaches from the late twentieth century and Cortesão's writings on indigenous knowledge from the 1940s. Taking into account the above questions, I divided the chapter into two sections. First, I will explore Cortesão's dialogues, considering how local debates and overseas references encouraged him to study indigenous people's spatial knowledge. Then, I will stress some textual and visual documents selected by Cortesão to discuss indigenous maps as specific categories. By following these steps, my primary intention is to discuss Cortesão's comments on indigenous mapping in the light of postcolonial and decolonial approaches, understanding exploration maps as co-produced and hybrid artefacts.

1.2 Cortesão's Dialogues: *Bandeirante* as a Myth and Indian as a "Live Map"

At first glance, Cortesão's historical narratives could be an easy target for postcolonial criticism. Connecting Portuguese expansionism with the exploration of Brazilian territory, Cortesão helped to develop the idea that there was a territorial "bond" inherited from the Portuguese that legitimized Brazilian borders (Oliveira 2013: 198). The harmonious and complementary encounter between indigenous people and explorers and the teleological discourse on the origins of Brazilian territory, clearly make Cortesão's writings susceptible to a postcolonial or decolonial critique. One can find statements where Cortesão opposes the "young age of the primitive indigenous natives" with the Portuguese, who "left behind Athens, Rome, and Jerusalem" (Cortesão 2009: 73), establishing essential features while classifying peoples through binaries that frequently served to colonialist narratives.

Reproducing a "standard narrative" of exploration, where the history of expeditions becomes an individual drama, with the explorer as the principal character (Driver 2013: 420), Cortesão highlighted the protagonist's participation of explorers and diplomats in the definition of Brazilian borders. According to Francisco Roque de Oliveira (2014), one of the main results of Cortesão's 17 years of political exile[2] in Brazil were his publications on Alexandre de Gusmão, the diplomat, and António Raposo Tavares, the explorer, or *bandeirante*. The focus on the diplomat, who negotiated a fundamental treaty for Brazilian borders, the Treaty of Madrid, and the explorer, who helped to integrate the space on the ground, was used to reinforce a well-known narrative in the history of Brazilian borders. Diplomats guaranteed *de jure* what Portuguese explorers, or *bandeirantes*, occupied de facto.

Cortesão's historical narrative highlighted both the explorers of São Paulo (*bandeirantes*) as well as the Portuguese political project, incorporating São Paulo's elite claims of historical legitimacy in the national identity project endorsed by elites from Federal Government located at Rio de Janeiro. The impact of Cortesão's publications and his lectures in the

[2]Jaime Cortesão was exiled from Portugal after his actions against the 28th May 1926 coup that brought military rule to Portugal for a period of 48 years. Cortesão was part of an important group of writers with republican, socialist and democrat orientation connected to the journal *Seara Nova* (dos Santos 1993). In order to confront the new government of Portugal the group tried to initiate a revolutionary process in the city of Porto in 3rd February 1927, but the attempt failed and Jaime Cortesão escaped to Spain. From that moment on, Cortesão and his family transited into many places. The outbreak of the Spanish war drove them to Barcelona and later to France. The advance of the German Nazis over France pushed them to Biarritz and along with other colleagues they returned to Portugal to escape the war (Saraiva 1953: 52). On 20th October 1940, after four months in Portugal, Cortesão leaved Europe to Brazil, arriving in Rio de Janeiro with his family aboard the S/S Angola in late 1940.

Brazilian Ministry of External Affairs (*Itamaraty*) from 1945 to 1950 (Moser 2005; Olivera 2014b), made him a prominent scholar in Brazil, requested by intellectual and economic elites from various regions that disputed the centrality of national identity. As stressed by Oliveira (2013: 199), Cortesão witnessed several political moments in the country but managed to be invited by different intellectual communities. Few foreign authors in Brazil during the mid-1950s could have a book published in Rio de Janeiro with the mention of President Getúlio Vargas (1953) and an exhibition curated in São Paulo for the Fourth Centenary of the São Paulo Foundation one year later[3] (1954) (Ribeiro 2018).

Contemporary analyses of Cortesão's historical narrative have shown how his research could be understood considering dialogues and mutual influences with historians and social scientists from São Paulo[4] (Oliveira 2013; Ribeiro 2018). In the decades preceding the arrival of Cortesão in Brazil, an influential intellectual elite worked intensely in São Paulo in the construction of local identity symbols, with the *bandeirante* as a founding myth. The works gathered in what could be called *bandeirismo* (Abud 1985; Raimundo 2004) influenced not only a research agenda in São Paulo but also the foundation of institutions, the naming of streets and the construction of monuments in honour of this general character named *bandeirante*. According to Raimundo (2004: 3), to these scholars, the "*bandeirante* was presented as the synthesis of São Paulo's spirit that constructed Brazil". Drawing upon previous researches on *bandeirantes* as Capistrano de Abreu's Colonial History Chapters [*Capítulos de História Colonial*] from 1907, authors such as Alcântara Machado, Basílio Magalhães, Afonso Taunay, and Alfredo Ellis Junior produced one new historiography of exploration during the 1920s and the 1930s (Raimundo 2004).

Afonso Taunay, for instance, worked since 1917 as a director of the São Paulo Museum [*Museu Paulista*], turning its exhibitions from a focus on nature to a historical account focused on the city of São Paulo and its relations with the conquest of the national territory. This shift was very emblematic for a museum founded on the margins of the river *Ipiranga*, where Emperor Dom Pedro I proclaimed Brazil's independence from Portugal (Raimundo 2004: 8). Taunay's book on the history of São Paulo's *bandeirantes*, published in 1934, presented a very standard narrative, constructing individual *bandeirantes* such as Raposo Tavares as a central character in the production of national identity. As Raimundo (2004) pointed out, the heroism of the *bandeirantes* eclipsed indigenous participation and the violence practiced in the colonization process. For the author, indigenous people did not receive the necessary attention in Taunay's historical narrative, and slavery seems fully justified as a means to enable territorial occupation. In fact, "in the historiography produced in São Paulo during the first decades of the twentieth century, the violent action of the *bandeiras* and its genocidal action of territorial invasion is seen as pure poetry" (Raimundo 2004: 8).

According to Raimundo (2004) the symbol of *bandeirante* was used with high intensity mainly during the commemoration of symbolic dates and creation of institutions in São Paulo. The commemorative exhibition of Brazil's 100 years of independence in 1922, the Constitutionalist Revolution of 1932, the foundation of the University of São Paulo in 1934, and the commemorations of the fourth centenary of the city in 1954 were some of these occasions of intense use of *bandeirante* as a symbol. These previous works undoubtedly influenced Cortesão,[5] and his

[3]The Fourth Centenary of São Paulo was celebrated with a series of events in 1954. The exhibition organized by Jaime Cortesão was opened to the public on September 14th 1954 in a new display space called *Oca* located at the *Ibirapuera* Park (Ribeiro 2018).

[4]Commenting on the networks that allowed Cortesão to be well accepted among intellectual elites from São Paulo, Ribeiro (2018:101) highlights the role of the Portuguese scholar João Sarmento Pimentel, who was based in São Paulo and helped to organize the first Cortesão's lecture in the city (Pimentel 1952).

[5]As Ribeiro (2018) pointed out, Taunay's influence on Cortesão could be studied by the letters exchanged among them. The São Paulo Museum director provided important

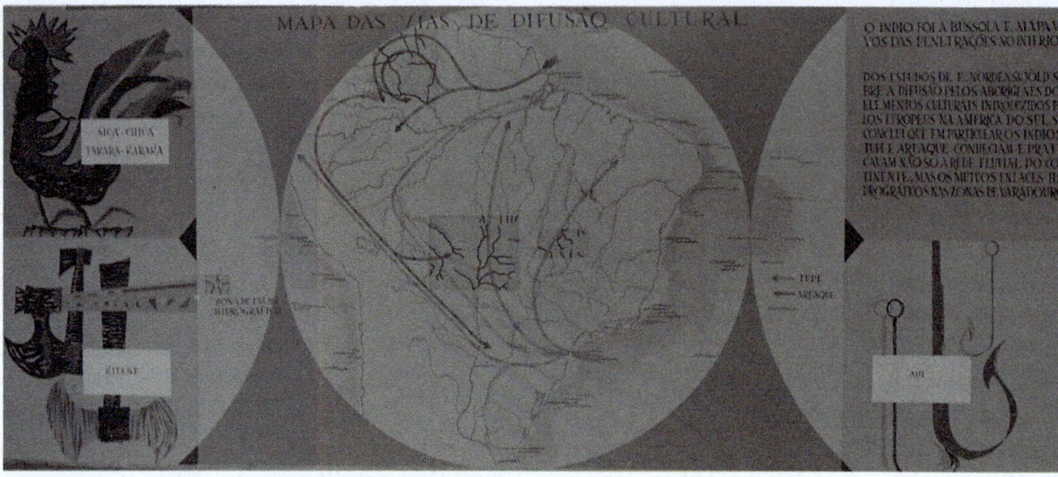

Fig. 1.1 Map displayed in panel 47 at the exhibition organized by Cortesão in 1954 during the celebrations for the fourth centenary of São Paulo city. National Library of Portugal. Expólio de Jaime Cortesão. BNP/ACP E25/1506

accounts met the demands of an elite from São Paulo who claim for centrality in the construction of the national identity. The name of his exhibition held in São Paulo in 1954 is representative of the political conciliation that the audience could find in Cortesão's work: "Historical Exhibition of São Paulo in Brazilian Historical Framework".

Praising the role of explorers in São Paulo's identity, Cortesão dedicated a whole section of his exhibition to *Bandeiras e Bandeirantes*,[6] showing the "origins, causes, and first consequences of the *bandeirismo*, an original and specific fact that defined São Paulo's history". In a panel drawn by Bernardo Marques,[7] the exhibition displayed the Portuguese origin and the military organization of *bandeirantes*.

Although endorsing many myths and heroes created by local historians, Cortesão also highlighted the role of indigenous knowledge in São Paulo's exploration. In the panel 47 of his exhibition[8] (Fig. 1.1), Cortesão placed a large map to show cultural diffusion circuits among *Tupis* and *Aruaques*. The text located on the right side of the map explained Cortesão's argument: "Indigenous people was the compass and the live maps in countryside penetration". The 1954 exhibition was not the first time that Cortesão used these expressions—"compass" and "live maps"—to describe indigenous knowledge. The subject appears in his talk given in 1944 at the auditorium of the Ministry of Education,[9] and he

information and references to Cortesão and also sent him a copy of his book on *bandeirantes* in 1949. See Taunay (1949) Correspondência a Jaime Cortesão, 4th August. 1949. BNP/ACPC. E25/1067, Portugal National Library.

[6]Document available at the Portugal National Library. Cortesão et al. (1954/55) Introdução. Document Op. cit. Section 4: Bandeiras e Bandeirantes. Espólio Jaime Cortesão. Arquivo de Cultura Portuguesa Contemporânea. Biblioteca Nacional de Portugal.

[7]Bernardo Loureiro Marques (1898–1962) was a Portuguese painter, illustrator, and graphic artist. He participated in the art renewal movement in the first decades of the twentieth century. In 1954 he collaborated in Jaime Cortesão's exhibition.

[8]The main documents consulted at the National Library of Portugal were the exhibition plan, written by Jaime Cortesão (Cortesão et al. (1954/55), BNP E25 Cx.64) and some photographs of the panels, mainly the photography of the panel 47 (BNP/ACPC E25/1506).

[9]Document available at the Portugal National Library. Cortesão (1944): A Cartografia antiga e os fundamentos geográficos e pré-históricos do Brasil. Conferência. Datil, 33 f. Rio de Janeiro, 12 set. BNP/ACPC E25/38. Biblioteca Nacional de Portugal.

used these expressions in 1947 in his newspaper essays published at *O Estado de São Paulo* and *A Manhã*.[10]

Indigenous agency in São Paulo and Brazilian history was certainly not a new topic among Brazilian scholars, and Cortesão found interlocutors on this subject among local historians.[11] However, unlike some of his predecessors, Cortesão seems to pay careful methodological attention to documents and theoretical references on indigenous agency in the exploration process. In his essay published in 1947, later reproduced in his book (Cortesão 1964), Cortesão aimed to stimulate one "new importance of indigenous participation in the *Bandeira*" (Cortesão 1964: 112). Cortesão seems aware that his approach to indigenous knowledge was relatively original in the historiography of Brazilian exploration. Respectfully, he "supposed" that "until nowadays there was not enough attention to this contribution" (Cortesão 1964: 104). By stressing the topographic sense of the indigenous people and their capacity to represent the territory, Cortesão highlighted the role of indigenous people as official "guides" or "living maps" in expeditions. To suggest a new research agenda on exploration, Cortesão wrote in the last paragraph of his essay: "it remains to be investigated to what extent this culture of orientation served to the *bandeiras* and influenced the first geographical and cartographic conceptions of the Portuguese" (Cortesão 1964: 112).

The map shown at the exhibition was also a simplification of an image previously presented in Cortesão's lectures. The map on "cultural dissemination routes" was first made by Ary Duarte[12] for Cortesão's classes on the work of the Swedish archaeologist and anthropologist Erland Nordenskiöld, who was quoted in the panel 47 of the exhibition. To challenge the idea that Portuguese explorers connected previously isolated indigenous areas, Cortesão detaily describes Nordenskiöld's investigations, who studied "in a remarkable essay" the diffusion of cultural artefacts introduced by the Portuguese among indigenous tribes in South America. According to Cortesão, the rapid dissemination of cultural elements "proves how fast and active communications across the continent were" (Cortesão 2009: 61).

To study the diffusion of "cultural elements" such as domestic fowls, bananas, and iron, Nordenskiöld used word mapping as an innovative methodology, composing a geographic information system with pre-computational techniques. The author's maps overlaid words given to these objects in various tribes and used transparent paper to locate the tribes in the territory. As commented by Cortesão (1964), the analysis of the maps shows that most of the cultural elements were dispersed through rivers, starting from the coast until reaching the Paraná River towards Paraguay and later reaching the Madeira system to the Amazon. Considering a "Guarani invasion of the Inca Empire", Nordenskiöld (1917) identified how words such as banana (paco, pacoba), needles (aui), knives (quise), and scissors (yetapa) circulated among the Guarani from Paraguay to Guyana. An exception to this pattern of circulation would be the domestic fowls, since, according to Nordenskiöld, these animals were spread by several routes and in an earlier period. Taking into account these references on indigenous circulation, Cortesão argued that "before the arrival of white people, a general language had begun to be created as an instrument of social and cultural unification" (Cortesão 2009: 69). The immense dissemination of certain

[10]The article entitled "Indian Compass and Live Map" [Indio: bússola e Mapa Vivo] was published by Jaime Cortesão in different Brazilian newspapers in October 1947. *A Manhã* published the text on 19 October 1947 and *O Estado de São Paulo* on 21 October 1947.

[11]Cortesão was also influenced by studies on indigenous agency in exploration carried out by Brazilian historians during the twentieth century, such as Sérgio Buarque de Holanda (Oliveira 2013). In addition, as Monteiro (2001) demonstrated, debates about the indigenous role in Brazilian identity and territorial formation were intense throughout the nineteenth century in institutions such as the Brazilian Historical and Geographical Institute [*Instituto Histórico e Geográfico Brasileiro*]. Although very relevant, these dialogues and influences are not scrutinized in this chapter.

[12]Ary Queirós Duarte (1905–1976) was a painter, engraver, and cartographer who collaborated with the elaboration of didactic cartograms for Cortesão's courses (Oliveira 2015).

post-Columbian words cannot be explained by the *lingua geral* (systematized by the Portuguese), "but only by the extensive migratory movements of Guaranis during the first part of the sixteenth century" (Cortesão 2009: 62).

The circulation through the territory enhances mapping skills, and Cortesão quoted references from different languages and academic backgrounds to highlight indigenous knowledge as a key aspect of exploration. The Russian geographer B. P. Adler, a "Ratzel disciple" (Cortesão 1964: 105), was an essential reference for his statements. In his 1910 book "Maps of Primitive Peoples", Adler exemplifies different cases of indigenous maps from all continents. A specific part is dedicated to the "Indians of South America", who are "not far behind the tribes of the northern part of the western world in map-making" (Hutorowicz 1911: 673). To exemplify the South American case, Adler quoted the German explorer Karl von den Steinen (1886), who reproduced in his book, *Durch Central-Brasilien*, some maps made by indigenous people from Xingu basin in Amazon. For von den Steinen, the indigenous capacities of spatial representation were well developed, especially considering that they had few previous contacts with "civilization". Drawing upon Adler's theories and von den Steinen's reports, Cortesão created one "law of the cartographic production", considering how the circulation within a vast territory had improved indigenous capacity for creating maps and figuring the best routes to circulate. Accordingly, the size of the Brazilian territory stimulates the spatial skills of indigenous peoples.

Cortesão's more in-depth studies on Nordenskiöld's and Adler's investigations show his intention in proposing a new research agenda, seeking to give "new importance to the participation of indigenous people in Portuguese geographical expansion" (Cortesão 1964: 104). Although in a different way from the perspectives of contemporary attempts to study the role of intermediaries as "guides, interpreters, porters and pilots in exploration" (Driver and Jones 2009), Cortesão (2009: 69) investigated how Indians provided "methods of protection, defence, and subsistence during walking marches". The natives transmitted "a proper technique of shipbuilding and river navigation, vast geographic knowledge of the territory and cartographic images that, although they represented a primitive art, helping the Portuguese expansion in the continent" (Cortesão 2009: 69). Taking into account these dialogues and references, one can consider what documents were selected by Cortesão to study the role of indigenous knowledge in territorial exploration.

1.3 Cortesão's Documents: Travel Writing and Maps as Sources on Indigenous Knowledge

In the first sentence of the opening essay of his book on the history of the *bandeiras*, Cortesão (1964: 11) presents a methodological challenge for scholars interested in the history of exploration: "there is no history without documents". According to him, this was an entirely accepted "truth" in different countries and academic communities. Cortesão, however, seeks to reverse this rule: "there are no documents without history", he states (Cortesão 1964: 11), arguing that all documents are actively embodied in a "system of ideas". These two "rules" seem to be very relevant in the study of the participation of indigenous knowledge in exploration. On the one hand, we must recognize the scarcity of documents and records on indigenous cartographic knowledge. On the other, it is possible to reinterpret historical materials in light of their history to produce new narratives about their meanings.

A similar methodological concern seems to draw the attention of contemporary scholars (Cusicanqui 2012; Mignolo 2012; Driver 2013). To investigate indigenous intermediaries as active agents as much as victims of European colonization, scholars have been looking inside diaries, reports, and travel writings. In addition to consult the explorer "own words to discover the part played by others" (Short 2009: 17), many research projects investigated maps, photographs, artefacts, and oral histories to reconstruct different hidden histories of exploration (Driver and Jones

2009). As well as many contemporary scholars, Cortesão discussed the challenge of the sources and highlighted the need to propose new readings of historical documents about exploration.

Although often associated with postcolonial and decolonial approaches, the strategy of investigating the contributions of Native Americans embedded in the explorer's narratives was applied in different contexts by Cortesão and many other Latin American scholars.[13] According to Cortesão, travellers' testimonies "allow us to conclude that they (the Indians) knew and transmitted, as a cultural heritage, a highly developed plastic art and cartographic representation of the territory". Cortesão identified, for instance, how Gabriel Soares described the *Tupinambás* as a people with "great knowledge of the land" in his Descriptive Treaty of Brazil" [*Tratado Descriptivo do Brasil*] from 1585. According to Soares, they "turn their faces to the sun, which serves as a guide, to circulate through places never travelled before" (Cortesão 1964: 107).

Cortesão also uses many reports of Jesuits as "proof" of the navigation skills of the indigenous. In his well-known 1654 letter written during his expedition to the Tocantins, priest Antônio Vieira detailed the construction of canoes and warned his reader. "Here it is important to note that the Indians make the canoes" and are "those who paddle them and often, as we shall see, those who carry them on" (Cortesão 1964: 115). Also, Cortesão was able to find eighteenth-century documents from the chamber of São Paulo highlighting how "without the brown people (Indians) one cannot make discoveries of gold" (Cortesão 2009: 51).

In addition to the natives' spatial abilities and navigation skills, some accounts are more directed to specific cartographic encounters. Cortesão describes, for instance, the reports by Alexandre Rodrigues Ferreira, whom he called the "Brazilian Humboldt". In his travel to Rio Branco, Ferreira wrote an episode when a *Macouxi* drew the river tributaries in the sand with a stick, and Ferreira invited him to reproduce the drawing with pen, ink, and paper. The Indian promptly attended the request, using a "series of angles more or less acute" to represent the mountain ranges and circles to represent the *malocas* houses in the tribes (Cortesão 2009: 111).

These cartographic encounters certainly helped Europeans in drawing their maps, and Cortesão investigated traces of indigenous information in such artefacts. By working at the National Library of Rio de Janeiro, Cortesão catalogued many unpublished documents, and his comments on these documents could offer some investigation paths for contemporary scholars. Taking into account Cortesão's academic and historical references, I would like to return to the group of maps from "Indians, *bandeirantes* and *sertanistas*" commented in the introduction of this chapter. One of the most outstanding of such documents was the *Mapa da região das moções de São Paulo a Cuiabá* (Fig. 1.2). Cortesão found this document divided into three pieces in different archive drawers, but the "stylistic identity" and the "continuity of the represented territories" were signs followed by Cortesão (2009: 232) to gather different parts of the map.

Cortesão stated that this was the first map of the *monções* region with the indication of rivers and paths and it was probably made around 1720. The map shows two large rivers and the Paraná River (*Rio Grande*) has its tributaries named with indigenous toponyms such as *Ivinhema* and *Uasuriú*. The only villages named are *São Paulo*, *Nazaré*, *Jundiaí*, *Parnaíba* and *Sorocaba* and all are placed in the right corner of the map. No villages are mapped along the way, and only the mines (*Minas)* are indicated in *Cuiabá* River (detail). The hybridisms between Indians and newcomer's information are explicit in this document because many place names are shown in two languages and distances are measured both by days of walking as well as in leagues (detail). By considering the region

[13]Monteiro (2001) draws attention to a Latin American and Latin Americanist historiography that, since the 1960s, sought to document and interpret the experience of native populations during colonization. Authors such as Leon-Portilla (1961) and Gibson (1964) explored chronicles of explorers as well as native documents such as manuscripts, testimonies, criminal investigations, and pictorial representations.

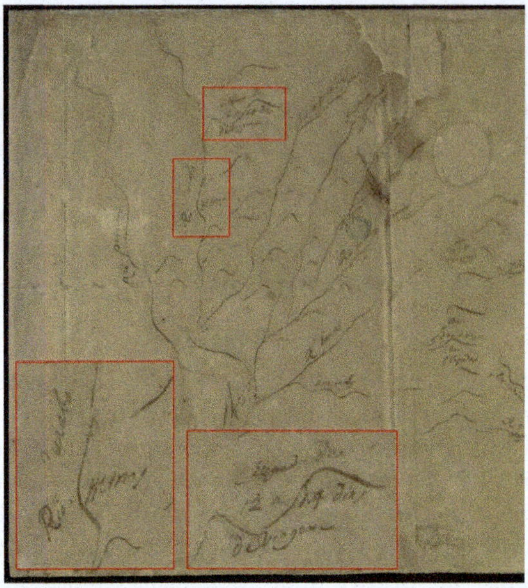

Fig. 1.2 Mapa da região das moções de São Paulo a Cuiabá ca. 1720. National Library of Rio de Janeiro. 55 × 104,5. BN Digital. Link to the document: http://acervo.bndigital.bn.br/sophia/index.asp?codigo_sophia=1472

represented, the focus in the mines, the information for navigation, and the influence of indigenous maps, Cortesão classified this map as a *bandeirante map*.

Cortesão used the cartographic language to differentiate *sertanistas* and *bandeirantes* maps, affirming that the maps of *bandeirantes* would be primarily utilitarian. In contrast, the cartography by *sertanistas* could be associated with "wise cartography", applying practices considered as "scientific" in European cartography. The fact that the maps are "primitive", "schematic", "limited to the hydrographic network" and made "essentially to be used in the field" (Cortesão 2009: 232) could, according to Cortesão, show the significant similarities among maps produced by indigenous and *bandeirantes.* To support his hypothesis, Cortesão suggested placing indigenous maps, such as those by von den Steinen (Fig. 1.3), side by side with *bandeirantes* maps, to show its "evident similarity" (Fig. 1.4).

In both maps, rivers were represented by lines and waterfalls by short "lines crossing them" (von den Steinen 1886: 214). Looking to the same maps commented by Cortesão, the Brazilian historian Sergio Buarque de Holanda (1949) highlighted how indigenous representational practices privileged communication over *mimesis*. In the same way of contemporary subway or touristic maps, indigenous practices often "free themselves from the pure visual image", because they do not intend to "reproduce all the sinuosity of the river course, but merely to indicate its extreme irregularity and warn the inexperienced traveller" (Holanda 1949: 183). The use of these simple cartographic practices could suggest how the "informative" and "memorial" role of indigenous maps were essential for the success of exploration (Holanda 1949: 183).

Although Cortesão had points of divergence and friendly competition with Holanda,[14] both

[14]Cortesão was recognized and accepted by São Paulo's intellectual elites, but he also faced resistance due to his conciliatory and Portuguese narratives. Discussing the historiography of Brazilian colonization, Reis (2007) identifies two groups, the authors who "rediscovered Brazil" as Capistrano de Abreu and Sérgio Buarque de Holanda and the authors who praised Portuguese action, such as Varnhagen and Gilberto Freyre. Cortesão sought to reconcile these perspectives, but was questioned for highlighting an intentional Portuguese action over Brazilian territory. Sergio Buarque de Holanda, for instance, wrote articles in the newspaper *Folha da Manhã* challenging Cortesão's thesis about the Portuguese explicit geographical state reasoning (Ribeiro 2018: 94).

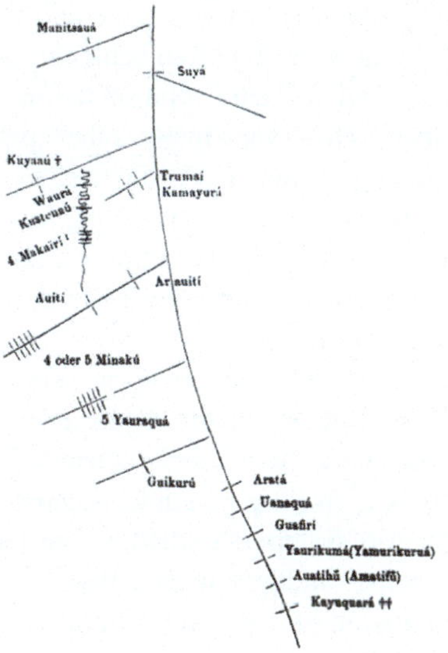

Fig. 1.3 Indigenous map reproduced by von den Steinen (1886: 214).

Fig. 1.4 Bandeirantes Map. Detail from Mapa com os três traçados diferentes do rio São Francisco. National Library of Rio de Janeiro. BN Digital. 29 _ 40.5. Link for the document http://acervo.bndigital.bn.br/sophia/index.asp?codigo_sophia=1705

authors agreed that indigenous people were developing a proper cartographic language. Cortesão (2009: 234) concludes that at the same time that a "scientific" revolution occurred in Portuguese cartography, a "native art of cartography" was created in Brazil, "more specifically in São Paulo". This cartography could be "at the same time simple, rude and vigorous to express the geographical facts in their utilitarian essentiality" (Cortesão 2009: 233). Taking into account these circulations of cartographic practices, Cortesão (2009: 234) suggested that the main characteristic of *bandeirismo* was "hybridism", "both in blood and in spatial cultures". European explorers drew upon native maps producing a complex juxtaposition in representational practices. This statement could denaturalize the "indigenous maps" as a fixed-category, understanding images made during territorial exploration as unfinished objects, drawn collectively by many hands.

Cortesão's references and documents are very pertinent to think about exploration maps as hybrid documents. Nevertheless, we should never lose sight of the fact that his use of the term "hybridism" also serves to produce a friendly and colonialist narrative. The creation of a new *Luso-Tupi* culture was allowed by Indians' qualities such as "tolerance" and "malleability" (Ribeiro 2018). It is precisely by similar political uses that the notion of "hybridity" is criticized by some decolonial scholars, such as Cusicanqui (2015: 94). The Bolivian sociologist identified a standard version of *mestizaje* in Latin American historical narratives, where the hybrid is presented as a new plant. This plant, however, is "based on a set of diverse roots", all subsumed to a hierarchical and Eurocentric identity pyramid. The ultimate end of this tactic has erased the memory of the Indian and "confine his remains in the museums", as an archaic "root" of a remote past (Cusicanqui 2015: 94).

Indeed, Cortesão stated that after the arrival of "mathematical priests", such as Diogo Soares, these hybrid maps were used to produce new images, obscuring their indigenous heritage by following the European mapping standards at that time (Oliveira 2014b: 12). Taking into account this historical narrative, Cortesão's criteria to classify maps—made by *bandeirantes* and by *sertanistas*—should also be criticized. By using the indigenous influence as the main criterion to identify *bandeirantes* maps, Cortesão restores a duality between native and Portuguese cartographic practices and do not extend indigenous influence to subsequent Portuguese cartography. A clear border is delimited between the indigenous mapping and a more "wise style".

Nevertheless, some examples were subclassified by Cortesão as "*sertanista* blurs" and they looked very similar to those influenced by the indigenous cartography. In the catalogue he created, all maps that excluded double names for the rivers, placed locations with pictorial symbols such as houses and churches, and applied some stylistic features from European cartography, were classified as *sertanistas* (Cortesão 2009: 243). The *sertanistas* maps generally cover a wider area and use symbols of European cartography but the information represented was not necessarily more precise or accurate. These inconsistencies appear in the case of the map *Mapa do Rio de Janeiro e São Paulo* (Fig. 1.5), which was classified by Cortesão as *sertanista*. Although it presents errors, such as the shape and the location of *Tietê* River, it represents São Paulo using a church (detail) and reduces the number of indigenous toponomies. Some names are crossed out, and no place is named with two toponyms (detail).

Other *sertanista*'s maps were classified by Cortesão (2009: 243) as having a "perfect style" that "does not authorize its classification as *bandeirante*". It is the case of the map *Região das Minas Gerais com uma parte dos caminhos de São Paulo e do Rio de Janeiro para as minas e dos afluentes terminais do S. Francisco* (Fig. 1.6), which presents a grid system and a scale in leagues. One should not neglect that the circulation of mapping practices from Europe allowed an improvement in the mapping of Brazilian territory. As Edney (2011: 332) argued, the challenge to conceptualize a "cartography without progress" should not be mistaken with simplistic statements such as "mapping has never progressed". In contrast, the author claims for a deeper understanding of how the word progress has been used and defined by the ideal of cartography (Edney 2019).

The adoption of leagues in scale and the grid system certainly contributed to achieving a more proportional and mimetic representation of the river courses. However, many other practices were still essential and were silenced in Cortesão's narrative on this map. In seeking to identify classificatory differences between *bandeirantes* and *sertanistas* maps, Cortesão neglects a series of continuities among them. Indigenous representational practices and territorial knowledge continue to compose many maps, even those made with a more "wise" and "scientific" style. Figure 1.6 is a striking example because while adopting a grid system classified as "perfect" by Cortesão, some "indigenous practices", such as the representation of waterfalls by cross lines are still being widely used (detail).

The practical function that privileges the communication of information, producing "schemes where everything aims at utility" (Cortesão 2009: 70) seems to coexist with a progressive normalization of maps that have been adapted to European stylistic standards. Among contributions and criticisms, the collected documents and texts written by Cortesão can provoke fundamental methodological debates for contemporary researchers interested in indigenous participation in the history of exploration.

1.4 Conclusion

The focus on knowledge exchange, considering how knowledge was appropriated as well as traded, could help to deconstruct fixed categories and open paths for new research agendas. Inspired by accounts that stress transits, translations, and communications in the process of knowledge production (Secord 2004; Raj 2006), scholars have been denaturalizing categories such as "local", "universal", "native", and "global". If we consider that European explorers drew upon native spatial knowledge and indigenous people drew using materials and information from the newcomers, we could develop a different gaze for "indigenous maps" as a category, considered it beyond a fixed and static cartographic genre.

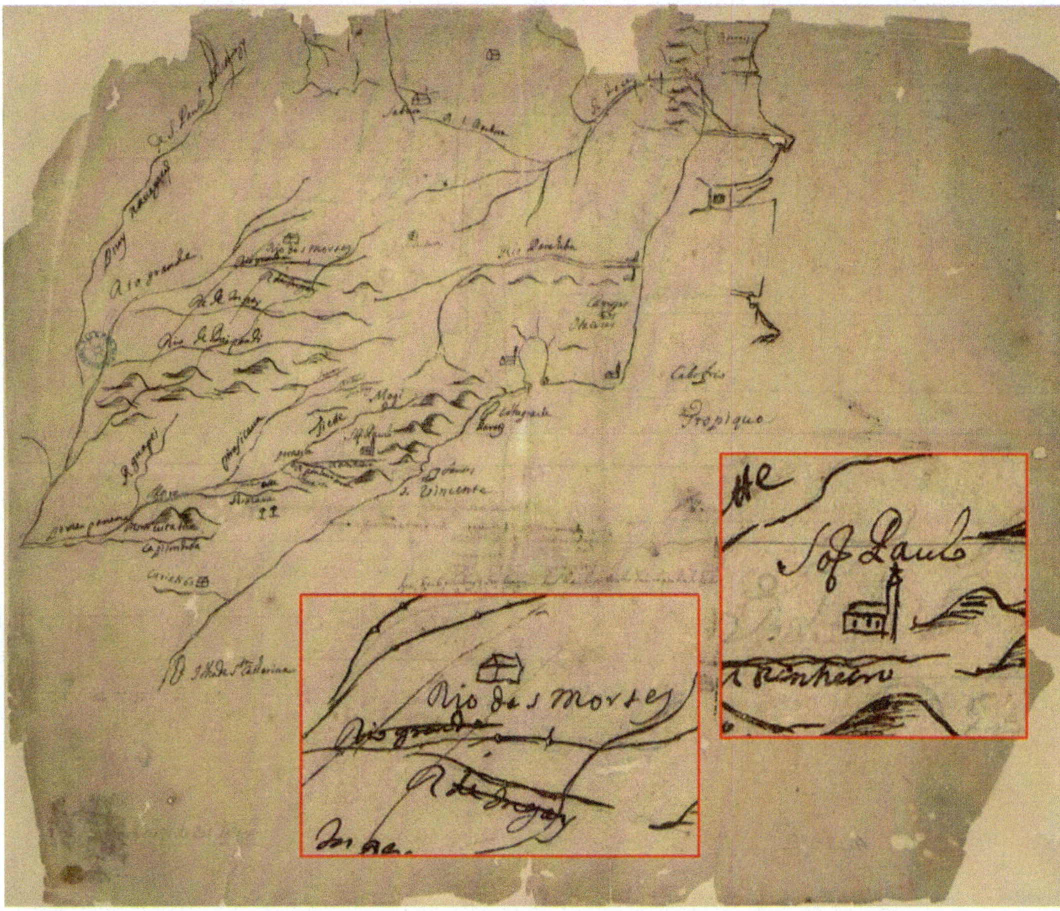

Fig. 1.5 Mapa do Rio de Janeiro e São Paulo. National Library of Rio de Janeiro. Mapa. BN Digital. 54 × 64. Link to the document: http://acervo.bndigital.bn.br/sophia/index.asp?codigo_sophia=1819

As has been shown in many investigations, the documents on the colonization of spaces and knowledge are often hybrids artefacts of cultural exchanges. By quoting studies from Lewis (1998) on native American mapping in the colonial period and Basset (1998) on West African maps, Driver (2016) suggested that the variety of materials and forms showed a much higher degree of hybridity than might be supposed (Driver 2016: 3). This hybridity raises fundamental questions on definition and approaches in the study of indigenous maps.

Cartographic encounters could be studied as cross-fertilization and constructive exchanges. Cortesão described, for instance, how the *Tupinambás* designated the Southern Cross Constellation as "*Crica*" or "*Curuça*", in *Tupi*, indigenous assimilation of the word *cruz*, showing how indigenous people were introduced to techniques of astronomical observation from the exchanges with the Portuguese. However, the study of these changes cannot overshadow power relations and hierarchies. Using the term symbiotic destruction, Short (2009: 12) is quite clear: "one culture wins, the other loses". The systematic use of indigenous slavery in the market of exploration shows how one group becomes dominant in the expense of others (Kok 2009). These power relations in knowledge production inspired some fundamental notion in contemporary decolonial approaches. In different ways, ideas on "coloniality of knowledge", "scientific

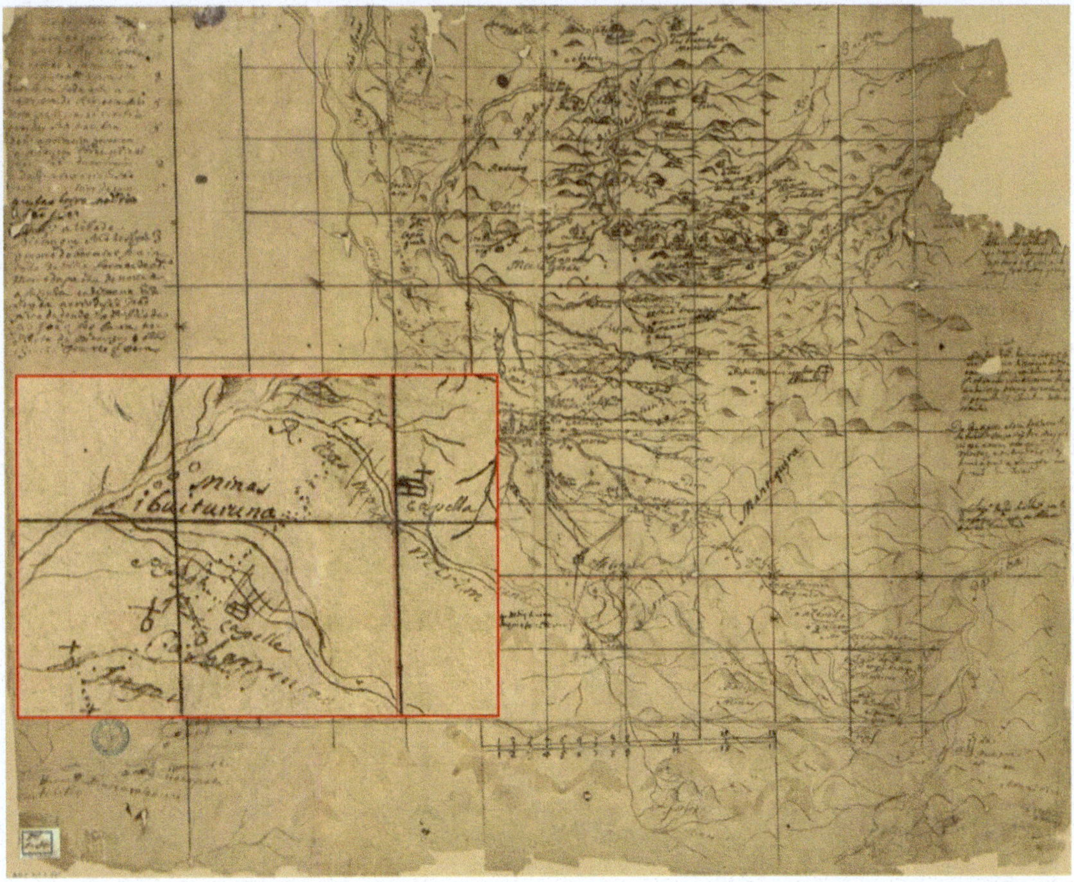

Fig. 1.6 Mapa da Região das Minas Gerais com uma Parte do Caminho de São Paulo e do Rio de Janeiro para as Minas, mostrando os afluentes do Rio São Francisco. National Library of Rio de Janeiro. BN Digital. 56 × 65.5. Link to the document: http://acervo.bndigital.bn.br/sophia/index.asp?codigo_sophia=1791

totalitarianism", and "epistemic violence" (Mignolo 2012), are all engaged on the consequences of colonialism in knowledge production.

Cortesão was undoubtedly far from any of these contemporary perspectives or criticisms. In his narratives, cartography is progressively purified, and after pages describing indigenous knowledge with enchantment, Cortesão (2009: 63) alerted: "when we talk about the indigenous geography, we move away, of course, from the idea of geographical sciences. It is a primitive culture, oblivious to the whole science of position and methods to accurately measure the space". Nevertheless, the debates and documents stressed by Cortesão to give "new light" for the indigenous participation in South America exploration, could meet contemporary demands on more profound dialogues about "indigenous maps" as a category. By exploring Cortesão's references on indigenous knowledge and the historical documents he selected to study this topic, this chapter aimed to recognize limitations and potentialities in Cortesão's writings for contemporary postcolonial/decolonial approaches engaging with "native" or "indigenous" maps.

On the one hand, Cortesão distinction between maps made by *sertanistas* and made by *bandeirantes* could reproduce a standard narrative of exploration. Although indigenous knowledge is recognized as necessary in early explorations, it is systematically erased, replaced, and incorporated, taking part in the prehistory of scientific explorations. On the other hand, by using indigenous influence as a fundamental feature for grouping maps made by *bandeirantes*, Cortesão also destabilizes personalistic and triumphalist narratives of exploration, which gives a secondary or accessory role to indigenous knowledge. The indigenous map is embedded in *bandeirantes*' spatial culture, and it is constitutive of the normative spatial knowledge produced in the history of explorations. Cortesão certainly presents a celebratory and harmonic view of knowledge exchange. But the presence of this topic in newspaper articles, books, and exhibitions demonstrates that indigenous maps were a subject of great interest for the author. Thus, Cortesão's writings may inspire future research agendas, considering exploration maps as hybrid documents and avoiding fixed categories that could restrict investigations on cross-cultural exchanges of geographical knowledge.

References

Abud K (1985) O Sangue Intimorato e as Nobilíssimas Tradições – A Construção de um Símbolo Paulista: O Bandeirante. Ph.D. Thesis, University of São Paulo, FFLCH—USP, São Paulo

Basset T (1998) Indigenous mapmaking in intertropical Africa. In: Woodward D, Lewis MG (eds) The history of cartography, vol 2, book 3. University of Chicago Press, Chicago, pp 24–48

Brückner M (ed) (2011) Early American cartographies. University of North Carolina Press, Chapel Hill

Cortesão J (1964) [1947–1949] Introdução à História das Bandeiras, vol 1. Portugália Editora, Lisboa

Cortesão J (1966) [1958] Raposo Tavares e a Formação Territorial do Brasil, vol 1. Portugália Editora, Lisboa

Cortesão J (1984) [1952] Alexandre de Gusmão e o Tratado de Madrid, vol 1. Livros Horizonte, Lisboa

Cortesão J (2009) [1957 and 1971] História do Brasil nos Velhos Mapas. Imprensa Nacional-Casa da Moeda, Lisboa

Cusicanqui SR (2012) Ch'ixinakax utxiwa: a reflection on the practices and discourses of decolonization. S Atlantic Q 111:95–109

Cusicanqui SR (2015) Sociología de la imagen: ensayos. Tinta Limón, Ciudad Autónoma de Buenos

dos Santos AR (1993) Jaime Cortesão: um dos grandes de Portugal. Fundação Engenheiro Antonio de Almeida, Porto

Driver F, Jones L (2009) Hidden histories of exploration. Researching the RGS-IBG Collections, Royal Holloway and RGS, London

Driver F (2013) Hidden histories made visible? Reflections on a geographical exhibition. Trans Inst Br Geogr 38(3):420–435

Driver F (2016) The indigenous map: native information, ethnographic object, artefact of encounter. Science Museum and Archives Consortium. www.royalholloway.ac.uk/geography/documents/pdf/newsarticles/project-details-indigenous-map.pdf

Edney MH (1993) Cartography without 'Progress'. Reinterpreting the nature and historical development of mapmaking. Cartographica 30(2–3):54–68

Edney MH (2011) Progress and the nature of cartography. In: Dodge M (ed) Classics in cartography. Reflections on influential articles from Cartographica. Wiley-Blackwell, Hoboken, pp 331–342

Edney MH (2019) Cartography: the ideal and its history. University of Chicago Press, Chicago and London

Gibson C (1964) The Aztecs under Spanish rule: a history of the Indians of the valley of Mexico, 1519–1810. Stanford University Press, Stanford

Holanda SB (1949) Índios e mamelucos na expansão paulista. Anais do Museu Paulista, Separata 13:177–290

Hutorowicz H (1911) Maps of primitive peoples. Bull Am Geogr Soc 43(9):669–679

Konishi S, Nugent M, Shellam T (2015) Exploration archives and indigenous histories. An introduction. In: Konishi S et al (eds) Indigenous intermediaries. New perspectives on exploration archives. Australian National University Press, Acton, pp 1–11

Kok G (2009) Vestígios indígenas na cartografia do sertão da América portuguesa. Anais do Museu Paulista, N. Sér 17(2):91–109

Lefebvre C, Surun I (2008) Exploration et transferts de savoir: deux cartes produites par des Africains au début du 19e siècle. Mappemonde 92(4):1–24

Leon-Portilla M (1961) The ancient Mexicans through their stories and songs. Fondo de Cultura, Mexico

Lewis MG (1998) Maps, mapmaking and map use by native North Americans. In: Woodward D, Lewis MG (eds) The history of cartography, vol 2, book 3. University of Chicago Press, Chicago, pp 51–182

Mignolo WD (2012) Local histories/global designs. Coloniality, subaltern knowledges and border thinking. Princeton University Press, Princeton

Monteiro, JM (2001) Tupis, Tapuias e historiadores: estudo de história indígena e do indinismo. Ph.D. thesis, Instituto de Filosofia e Ciências Humanas, Universidade Estadual de Campinas, Campinas

Moser HR (2005) The history of cartography in Brazil in the 1940s: Jaime Cortesão's lecture courses. Imago Mundi 57(1):70–74

Nordenskiöld E (1917) The Guarani invasion of the Inca empire in the sixteenth century: an historical Indian migration. Geogr Rev 4(2):103–121

Nordenskiöld E (1922) Deductions suggested by the geographical distribution of some post-Columbian words used by the Indians of South America. Erlander, Göteborg

Oliveira FR (2014a) Jaime Cortesão no Itamaraty: Os Cursos de História da Cartografia e da Formação Territorial do Brasil de 1944–1950. Scripta Nova 18 (463):463–499

Oliveira FR (2015) História da cartografia brasileira e mapoteconomia segundo Jaime Cortesão. Terra Brasilis (Nova Série) 4. https://doi.org/10.4000/terrabrasilis.1108

Oliveira TK (2013) Cartografias do sertão: os mapas sertanistas no discurso histórico de Jaime Cortesão e Sergio Buarque de Holanda. Territórios e Fronteiras (UFMT. Impresso) 6:188–210

Oliveira TK (2014b) Cartografia e Natureza: os mapas do sertanismo (primeira metade do século XVIII). Revista Eletrônica Documento/Monumento 13:9–24

Pimentel JS (1952) Portugueses emigrados políticos no Brasil. Seara Nova, Lisboa

Raimundo SL (2004) Bandeirantismo e identidade nacional. Terra Brasilis 6:1–18

Raj K (2006) Relocating modern science. Circulation and the construction of scientific knowledge in South Asia and Europe. Palgrave Macmillan, Basingstoke

Reis JC (2007) As identidades do Brasil: de Varnhagen a FHC. Fundação Getúlio Vargas, Rio de Janeiro

Ribeiro DWA (2018) "São Paulo capital geográfica do Brasil": a exposição do IV Centenário de São Paulo e a formação do território brasileiro na escrita histórica de Jaime Cortesão (1940–1960). Intermeios, Fapesp, São Paulo

Saraiva R (1953) Jaime Cortesão. Subsidios para a sua Biografia, Seara Nova, Lisboa

Secord JA (2004) Knowledge in transit. Isis 95(4): 654–672

Short JR (2009) Cartographic encounters. Indigenous peoples and the exploration of the New World. Reaktion Books, London

Taunay A (1924–1950) História Geral das Bandeiras Paulistas, vol 10. Imprensa Oficial do Estado, São Paulo

von den Steinen K (1886) Durch Central-Brasilien. Expedition. Zur Erforschung des Schingú. F. A. Brockhaus, Leipzig

2 Pioneers of Latin American Critical Geography: Josué de Castro and Antonio Núñez Jiménez

María Verónica Ibarra García and
Edgar Talledos Sánchez

Abstract

This work aims at providing a regional look at the origins of the critical discourse within Latin American geography. It discusses the contributions made by two authors of the region, who worked on geography identifying and analyzing conditions of poverty, marginalization, lack of democracy, and denouncing the power holders who, consequently, were responsible for the inequality conditions that deprived the region. Under a critical look, several proposals were developed, proposals which exposed imperial, dependency and unequal relationships in Latin America. That is the case with the geographical writings by Josué de Castro and Antonio Núñez Jiménez, who, through their work, marked a milestone in territorial, regional and spatial knowledge of Brazil, Cuba and Latin America.

M. V. Ibarra García (✉)
Facultad de Filosofía y Letras, Colegio de Geografía, Universidad Nacional Autónoma de México, Distrito Federal, Mexico
e-mail: maveroibar@gmail.com

E. Talledos Sánchez
CONACYT/El Colegio de San Luis A.C., San Luis Potosi, Mexico
e-mail: edgartalsan@gmail.com

Résumé

Ce chapitre vise à établir un regard régional sur les origines du discours critique dans la géographie latino-américaine. Cela contribue à la reconnaissance des contributions réalisées par deux auteurs de la région qui travaillèrent des thèmes de la géographie en identifiant et analysant ses conditions de pauvreté, marginalisation, manque de démocratie. Ils dénoncèrent notamment les détenteurs du pouvoir y pour cela les responsables des conditions d'inégalité qui sévissaient dans la région. A travers d'un regard critique, on développa des propositions qui exposèrent les relations impériales, inégales et dépendantes de l'Amérique latine. C'est le cas des textes géographique de Josué de Castro et Antonio Núñez Jiménez, lesquels, par leurs travaux, marquèrent une distance et une ligne de discussion sur la connaissance territoriale, régionale et spatiale du Brésil, de Cuba et de l'Amérique latine

Resumen

Este trabajo tiene por objetivo brindar una mirada regional a los orígenes del discurso crítico dentro de la geografía Latinoamericana. Contribuye al reconocimiento de las aportaciones que realizaron dos autores de la región que trabajaron temas de la geografía e identificaron y analizaron las condiciones de

B. Schelhaas et al. (eds.), *Decolonising and Internationalising Geography*, Historical Geography and Geosciences, https://doi.org/10.1007/978-3-030-49516-9_2

pobreza, marginación, falta de democracia, así como a la denuncia de los detentores del poder y, por ende, de los responsables de las condiciones de desigualdad que privaban en la región. Bajo una mirada crítica se desarrollaron propuestas que exhibieron las relaciones imperiales, de dependencia y desiguales en América Latina. Este es el caso de los textos geográficos de Josué de Castro y Antonio Núñez Jiménez, que por medio de sus obras marcaron una distancia y línea de discusión sobre el conocimiento territorial, regional y espacial de Brasil, Cuba y de América Latina.

Keywords

Latin American Geography · Critical geography · Josué de Castro · Antonio Núñez Jiménez

Mots-clés

Géographie Latino-Américaine · Géographie critique · Josué de Castro · Antonio Núñez Jiménez

Palabras Clave

Geografía Latinoamericana · Geografía crítica · Josué de Castro · Antonio Núñez Jiménez

2.1 Introduction

Throughout the whole second half of the nineteenth century and the first half of the twentieth century, geographical thinking revolved around the aegis of what was discussed in universities of France, Germany and the United States. During this period, there was a plurality of geography definitions and political, theoretical and epistemological stances which included positivism, historicism and Marxism. Therein, the political-historical events of the First and Second World Wars, the interwar period, the 1929 crisis and Cold War, were relevant events which stimulated discussions within geography. From the end of the nineteenth century, this latter had already been defined as an academic discipline.

In these discussions, the geographers elaborated concepts of space, region, territory and landscape which were crucial for the theoretical structuring developed during the long journey of the geographical discourses. This may be exemplified by works of Paul Vidal de La Blache, who defined geography as the study of regions; of Alfred Hettner, who conceptualized it as the differentiation of areas; and of Carl Sauer, who considered it as the study of humanized landscapes, which later would give way to cultural geography (Moreira 2007: 64).

In spite of this plead of geography's definitions, which spread from European and American universities, proposals were also made by universities of Latin America and the Caribbean, mainly inspired by anarchism, Marxism, structuralism, phenomenology, to mention but a few. Here, the communication among geographers from Europe, the United States and Latin America, was constant; the dialogues established between them were not mechanical or coincidental; they were marked by the same historical events. Suffice it to see the debates that Isaiah Bowman and Jean Bruhnes had at the Paris Conference in 1919,[1] or the polemics arisen among a wide assortment of geographers at the 18th International Geographical Congress at Rio de Janeiro in 1956 (De Bomfim and Mello 2012), where according to Ángel Bassols-Batalla (1925–2012), there were:

> [...] about 220 international, effective delegates came from the 5 continents and 56 countries, from Norway to South Africa, Canada and Chile, from India to the Australian Union, Egypt to Japan. Along with countries of old geographical tradition, such as France, Germany and the United States, representatives of new nations, such as Ethiopia. Iraq, Iceland, Rumania and Pakistan, went to Rio. Even delegates of English and Portuguese colonial dependencies were present, Gold Coast and Angola, for instance, and former French territories

[1]As an example of this, it is necessary to remember the debates that took place at the Peace Conference in Paris, 1919, where geographers attended: Isaiah Bowman, Jean Brunhes, Emmanuel de Martonne, Albert Demangeon and Alan G. Ogilvie (Geoffrey 1980).

(Morocco and Tunisia). With the participation of the Soviet Union, the only non-represented great country was China, due perhaps to the acceptance, on behalf of the Congress, of credentials presented by geographers from the rival government in Formosa. There were also observers of the United Nations, -UNESCO and other international organizations. From Latin America, 13 countries were represented, including the Island of Puerto Rico (Bassols 1957: 62).

All these meetings were characterized by large discussions within the geographers' community on what has been considered as geographic in different historical stages and within those periods. That is what makes Souza (2006) state that: "For each society its space-time, or its characteristic spatiality" (p. 28).

In this sense, geography had little to do with the conservative, colonial and essentialist version which considers geography as part of the physical, 'natural' sciences, or as the description of valleys, rivers, mountains, volcanoes, plains, to be learnt by memorizing and positioning data and names represented in a map. As Pierre George noticed, "in a landscape [some scholars] only come to see quarries to take samples from, to make them undergo the study or refrigerator trial, to winnow them, to establish the grain size chart, to note down the hardness index, to prepare thin layers and examine them under the microscope with polarized light, etc." (George 1969: 6).

Studies on geography and notions on 'the geographic' passed gradually through long controversies between different theoretical approaches. As it may be observed in the large tradition of theoretical and political dissidence in geography, as well as in Élisée Reclus and Piotr Kropotkin proposals from anarchism: "Geography is but history in space, just as history is geography in time," wrote Reclus (1986: 339). Besides, he considered that: "Social Geography should reveal class struggle, the search for balance and the individual's sovereign decision" (Reclus 1994: 219). In this way, a plurality of leftist traditions converged in the investigations carried out by geographers, prior to the 'critical turn' at the end of the Sixties and Seventies of the twentieth century.

In the first half of the twentieth century, the political contexts were set by Roosevelt's *New Deal*, Italian fascism, German Nazism, Latin American totalitarianism, while decolonization and Arab nationalist regimes determined politics intended as a war against the communist and anarchist left stances (Fermandois 2016: 96). It was during this stretch that Josué de Castro (1908–1973) and Antonio Núñez Jiménez (1923–1998) stood up as critical voices denouncing the social conditions in their countries. Their arguments anticipated the critical movement of geography renovation, during the 1960s in the United States,[2] France and Brazil, which called into question its classic discourse and the *New Geography's* neo-positivist postulates of the 1950s. Both authors presented proposals that have represented visions which are different from the conservative ones, inside and outside geography.

2.2 Pioneers in the 1940s and 1950s

Even though de Castro and Nuñez Jiménez are not usually included in the most famous streams of critical and radical geographies in the Sixties and Seventies, they were two figures who made convincing critiques to the political and economic conditions of dependence that explained poverty, socio-spatial marginalization and hunger in Latin America and in the world. This

[2]In August, 1969, a group of academics and students of Clark University founded the journal *Antipode. A Radical Journal of Geography*. It aimed at making a critique of the status quo and the conservative positions of geography and social sciences at that time. Its purpose was not just the academic critique, but the social transformation following the ideals of social justice through radical change. In this way, the founders of the magazine, Ben Wisner, David Stea, J. Richard Peet, Richard Morrill, J. M. Blaut, Karen Thompson, Bill Emerson, Fred Donaldson, Reed F. Stewart, Ruby Jarrett, Robert W. Kates and Jeremy Anderson stressed the necessity of being radical in the etymological sense of the word. Linguistically, this comes from the Latin word radix, which means 'root', to reveal the essence of the phenomena that would allow presenting alternatives of change is society (Castree et al. 2010: 3–4).

chapter analyzes historically some of their most famous books, namely, de Castro's *The Geography of Hunger* (1946) and *Geopolitics of Hunger* (1951) and Nuñez Jiménez's *Geography of Cuba* (1954).

It is in this context that critical geographies were developed. Although critical approaches were not common in Latin American social sciences in the Fifties, there have been pale approaches between geography and a still unconventional Marxism. The possible military intervention of the United States in Cuba was also denounced by discussing Cuba's specific territorial conditions (Núñez 1960). It is worth noting that this text is only an introductory piece. These authors' works should be analyzed with greater profundity in the future. For the time being, we will present these two Latin American social scientists who published work that should be considered as a major reference on how a geography with explicit positionings against socio-spatial injustices, did exist.

These geographic works placed 'science' in a position different from the one traditionally assigned, which excluded critiques of the social, political or economic conditions that thrived in each of the political spaces of their respective nations.

Our journey begins in the 1940s with Josué de Castro, the author of two pioneering and very potent works *Geografia da fome* [*Geography of Hunger*] (1946), and subsequently, *Geopolítica da fome* [*Geopolitics of Hunger*] (1951). Then, we present the case of Antonio Núñez Jiménez, with his work *Geografía de Cuba* [*Geography of Cuba*] (1954). Here, one can appreciate two pioneers who frontally questioned the life conditions of their countries. While they may be considered rather 'pioneers' than fully flagged 'critical geographers', these authors took clearly political positions in geography and through geography, challenging the determinist, neo-Malthusian, apolitical and neutral stances that quantitative geography proclaimed in the Fifties.

The book *Geography of Hunger* of 1946, by Josué de Castro, is of the highest importance. It was a book of wide national and international circulation, which generated an impact beyond geography and Brazil.[3] He was not always considered as a geographer, if not as a medical doctor. Yet, he was professor of geography at the Federal University of Rio de Janeiro. "In 1939, he joined the recently created Philosophy, Science and Letters School of the Brazil University-currently Federal University of Rio de Janeiro. He occupied the Chair for Human Geography, by mean contest in 1947, with the thesis 'City of Recife, an Urban Geography Essay" (Linhares Leite 1984: 333). In the work that concerns us, he explicitly sustains that it will be through geography that the study will be carried out and cites several geographers, who guide his study.

> We decided to face the problem about a new perspective on a more distant plane, where a panoramic view may be obtained overall; a vision where some small details, certainly, dimmed off, but in which relationships, influences and multiple-factor connections stand out in a comprehensive way that intervenes in the manifestation of the phenomenon. For this purpose, we intend to launch the geographic method, not the study of the hunger phenomenon. It is the only method that, from our point of view, allows us to study the problem in its whole reality, without taking away the roots that bound it underground to other innumerable economic and social manifestations of life of the peoples. It is not the descriptive method of ancient Geography, it is more the imperative method of the modern geographic science, which is embodied within the fecund thoughts of Ritter, Humboldt, Jean Brunhes, Vidal de la Blache, Griffith Taylor and so many others (De Castro 1984: 34).

In this sense, we may consider that this text is still inside a reading of traditional geography, in the possibilist approach. Nevertheless, these authors presented several elements of originality; one of them is that de Castro's book did not coincide with the traditional regional or national

[3]Books which practically produced a great impact in all the countries of the world, which is why they were translated into 24 languages in consecutive edition. Being the first time, in the international public opinion, it alerted on the problem of hunger, a stigma of underdevelopment and residue of the social and economic structures inherited from colonialism (De Castro 1984: 334).

monographies and presented hunger as its thematic focus. Another element presented, and that turns out to be novel, was the critic toward plantations, monocultures and colonialism (the latter was responsible for the first two). This book had a great impact; it even sold out in a few months, which gave its author a great visibility.

Five years later, in 1951, de Castro's most important work *Geopolitics of Hunger* was published. It would become a model in the social sciences of Latin America and the world; besides, it represented an advance of great importance compared to *Geography of Hunger*, because it explicitly challenged the conservative naturalist positivism, which permeated geography and other sciences. In this sense, de Castro rejected the determinist proposals, identifying the hunger problem as a social, economic construction and disassociating it from the alleged physical-geographic conditions that determinist readings presented as the cause of the problem. Simultaneously, the author criticized Neo-Malthusianism, positioning himself in a clearly leftist or libertarian perspective.

The author discussed hunger conditions worldwide, arguing for the existence of hunger conditions of different magnitudes in every country. However, no part of the world was exempted from scrutiny, except for what in that moment was the USSR, given that de Castro declared not having any reliable sources on that case. Apart from that, four continents were analyzed: America, Europe, Africa and Asia.

> Hunger and economic imperialism. Regarding the hunger taboo, there were stronger reasons than the moral preconceptions. Reasons whose roots were buried in the remote world of privileges that always worked to hold back the exam of the hunger phenomenon in a modern, intellectual scene. Due to the economic imperialism and profit ambition, it was necessary that the production, distribution and consumption of edible products continued the process indefinitely, as pure economic phenomena guided by the feeling of their exclusive financial interests and not as phenomena of the highest social interest for the benefit of collectivity. Modern European civilization—which reached its peak due to the geographic horizon expansion of the world after the sixteenth century, and to the global economy that followed it—in a convenient way divulge, in its world of apparent splendor, the disgusting tragedy of hunger, product, to a great extent, of its dehumanized colonialism (De Castro 1957: 30).

Two elements of the highest importance—for the debates in the critical geography that he launched and for Brazilian geography—are doubtlessly de Castro's frequent references to geographers and geographical debates. In this way, in the text, we find references to the postulates of conservative geographers, such as Ratzel and Huntington, whose flawless determinism and racism were criticized by de Castro; as a consequence, the Brazilian author referenced Reclus's discussion on the problem of hunger in India in the *New Universal Geography* as early 'geographies of hunger'. De Castro argued that, although the problem had always existed, it worsened during the nineteenth century as a responsibility of the English politics in the economy of India and quoted the anarchist geographer: "In the last years of the same century," according to Reclus, "more than 20,000,000 people died of starvation and only in the terrible year of 1877, more than 400,000" (De Castro 1957: 191).

Another key element was the critique of latifundia and the subsequent plantation agriculture, which gave priority to the exportation of agricultural products, such as sugar, cotton and coffee, among others, over food production for local self-consumption. Therefore, de Castro considered the agrarian reform as an element of the highest importance to revert this situation; but at the same time, he was aware that not only food fair distribution is necessary. Indeed, also social and material conditions that make food independence possible needed to be considered, as it is showed when the author identifies the Agricultural Reform in Mexico as a step ahead against hunger, despite the challenges that land distribution without financial and technical resources might entail. "The revolutionary Mexicans, more idealist than technical, forgot that it was not enough to distribute land, but that it is precise to supply the technical resources to carry out their appropriate farming" (49).

On the other hand, the book by Antonio Núñez Jiménez, *Geography of Cuba*, written in 1954, was forbidden and its copies were burnt by order of Cuban dictator Fulgencio Batista, and very few works on this author exist outside Cuba. After the triumph of the Cuban Revolution, the book was reedited and it became the official book of the Cuban geography. Our analysis is based on the reedition of that book; some elements changed, which without a doubt, are already the result of a clear political positioning after the Revolution. However, others expose the tension between a traditional geography and a new geography in formation.

Another interesting element is that at being considered the official book of geography, Núñez Jiménez's work gave a paramount contribution to the formation of the Cuban geography, but, conversely, it did not enjoy great consideration in the rest of Latin America. Almost no text about critical geography cites or even mentions it, which also leads to important questionings on science and its relationship with power, as we discuss below.

The *Geography of Cuba* was still conceived as a 'classical' national monograph, organized in five parts:

First Part—Physical Geography
Second Part—Human and Political Geography
Third Part—Natural Resources
Fourth Part—Economic Geography
Fifth Part—Regional Geography.

As can be noted, it follows a monographic order, that of the classic geography. In general terms, the book describes, in an accurate way, the characteristics of each of the Cuban regions. Nevertheless, it presents interesting elements to consider, given that, as it was mentioned, it is the beginning of a new geography that is growing apart from the old paradigms that geography had been treating, such as determinism, as well as an alleged neutrality, clearly taking a political stance. In that sense, what Antonio Núñez Jiménez argued on page 21 is crucial:

> It is true that geographic context has an influence in the human and historical determinations, especially in the primitive stage of humanity. In present times, the terms have been inverted extraordinarily. Today, man mostly dominates his geographic environment. He can control a flooding: join two rivers as fast-flowing as Volga and Don; make the Antarctic inhabitable; turn gardens into the driest desserts. He turns dry weathers into humid ones; lowers temperature in hot places and produces heat in cold environments. Modern man not only changes the physical conditions but is also capable of modifying characteristics of his own being and correcting hereditary factors. (Núñez 1960: 21)

Further on, the author discards arguments by Ganivet, Montesquieu, Huntington and Francisco Bulnes, for coming from determinist origins "it would be idle to insist on the lack of scientific basis of those theories that have fallen by themselves." But there are still 'intellectuals' who pretend, every day:

> [...] the inferiority of Latin Americans due to the race blend, the physical environment or for other reasons. All those theories, without scientific basis, tend to justify exploitation, injustice, political and international abuse, in order to create misunderstandings and block our nations' progress... The economic and social regime must be blamed (held responsible) for many of the misfortunes or virtues that some try to explain through the weather, topography or sky clearness. (Núñez 1960: 22)

However, this radical introduction, in a geography text in the Fifties, was followed by ten chapters of monographic geography, describing each region from its geologic origin to political geography and population. It is only in the second part of the book that we find again elements of critical analysis, related to the issue of natural resources. The geographer explained that Cuba's full industrialization had not been achieved "because the great international monopolies that take advantage of our resources have opposed to it; they control our banks, the main public services, most of the public transportation, and the existing industries, such as land, mines, marine and aerial traffic, and the other means of wealth production" (Núñez 1960: 209).

An important element is Núñez's discussion on land concentration in the hands of foreign capital, as well as the subsequent demand for a land reform. It is actually this element that has been considered to cause Batista's disappointment and the reason why he ordered the disappearance of the book. This prickly geography contained information on the agricultural conditions in Cuba, exposing the miserable conditions in which the rural population lived; at the same time, it displayed numbers of the great, American sugar companies in Cuban territory, number of sugar mills, average capital, current capital, gross profits and obtained surplus. Data were also presented on the sugar latifundia, as well as for the grain and resource extraction (rice, coffee, minerals, forests); simultaneously, Núñez showed poverty and marginalization conditions of the Cuban people, whose logical consequence was the demand for the land reform. This subject encompasses most of the book's third part, dedicated to natural resources.

Geography is presented as a science that takes political position by evincing the inequality conditions that prevailed in Latin American regions. Even though, in Núñez's work, there is no explicit reference to Marxism, there is an explanation of the poverty conditions that are related to the political and economic conditions of the island and to US imperialism. In this sense, one may consider this a critical geography 'in becoming'.

A key point of Núñez's book is the study of the imposed by the US to Cuba, as well as the colonialist consequences of the Platt Amendment,[4] so the work aims to explain the dependency in the relationships between them. The author focuses on these relationships between countries. His vision is not deterministic and presents mechanics of annexationism versus dependence.

In this problematization, there are two other elements that must be considered; geography in Cuba, as the rest of the scientific reordering, was the result of the politics of the Education Reform in 1962. "The geographic discipline is assigned to geosciences. At the same time as Geology, the Geography School was founded in 1962, also within the Science School, it would obtain the quality of Geography School in 1978, when it was opened in the urbanization of Alamar, to the east outside the capitol. Since then, it articulates over two main branches: physical geography, stronger at the beginning, and economic geography, which has been developed little by little. The information technology equipment and the spatial analysis methods were essentially introduced by the socialist countries" (Dory and Douzant-Rosenfeld 1995: 65).

The same authors argue that, since 1970, the influence of schools that were traditionally recovered, the French, Spanish and North-American schools, was diminished by the influence of the Soviet geography (Dory and Douzant-Rosenfeld 1995), heavily relying on notions of Cartesian space, territorial planning and ordering.

Having described these considerations about the work of these two authors, one may find, within their work, coincidences of critique toward conservative discourses inside and outside geography; as example we may mention

1. The critique of determinism,
2. The critique large estates and monoculture,
3. The critique of colonialism and imperialism as it is responsible for the first two,
4. The importance of the Land Reform to resolve the worst problems, inequality and hunger.

In all this, we may consider these authors as pioneers of a critical look to social and economic conditions of their respective countries, including critiques of determinism, racism and naturalization of hunger and social inequalities, and at the same time showing the injustice represented by large estates and plantations, by the USA interventions and the capitalist companies.

Having said that, it is necessary to specify that Núñez Jiménez became an author of the triumphant regime and the Cuban revolution adopted the Cartesian readings of space then consolidated in the socialist countries. De Castro

[4]This was an appendix to the Constitution of Cuba of 1901, imposed by the United States, in order to regulate relations between the two republics. Its name came from the American senator Orville H. Platt, who wrote it. Platt presented the aforementioned amendment in the US House and was accepted as a law was imposed on Cuban under the threat the United States not vacating the Island (Macías 2001: 111).

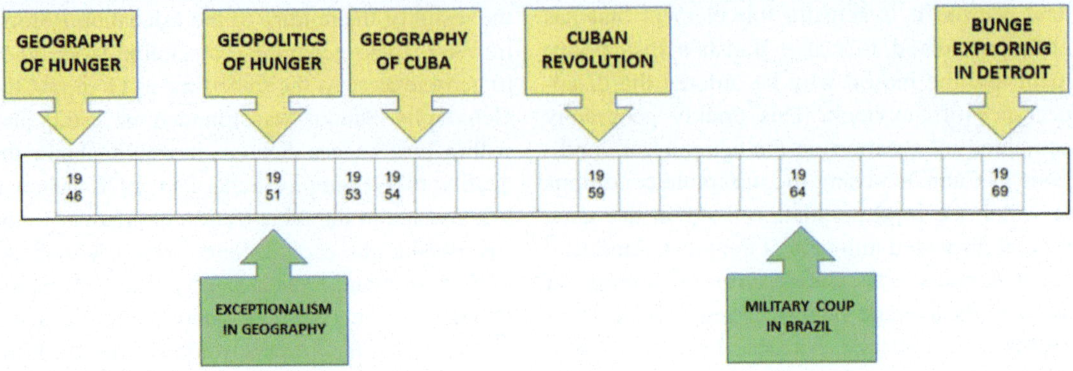

Source: Own elaboration, 2017.

Fig. 2.1 Context in which the first critical texts were presented in the Latin American region. *Source* Own elaboration, 2017

continued denouncing social inequalities and had to leave for exile when the Brazilian military seized the power in 1964. Still enjoying the great prestige that he earned worldwide with his scholarly critiques and political struggles, he died exiled in Paris in 1973.

A much more important element, at least for the case of Mexico, is the ignorance or oblivion into which these works have been cast. It seems as if they never existed and that the critical view of geography started with the English-speaking authors. What happened then? Where did this geography moved to? How much did it lay precedent to a Latin American critical geography? Unfortunately, it didn't have an impact on the elaboration of Latin America's critical geography. We have two hypotheses over which ignorance of these works travels, which initiate a look within geography, different from the traditional ones.

The first element of this hypothesis is that in the same years a fundamental work on geography that came to revolutionize geographic science, that is the text of German geographer exiled in the United States, Fred Schaeffer, *Exceptionalism in Geography* (1953). This work made a devastating critique of traditional geography as synthesis or 'ideographic' science, and proposed the 'nomothetic' recovering of the concept of Cartesian space. Right in that time of the rising of quantitative geography, these Latin American geographers published works that went against apolitical and neutral visions of geographical sciences.

On the other hand, Cuban geography, as it was mentioned, has a contradictory movement; on the one hand, an antiimperialist, anticapitalistic rhetoric; on the other, it assumed the principles of Cartesian Space from socialist geography.

One element that cannot be left out and that adds to the complexity is the political context. In 1964, the Brazilian military coup marginalized all the positions of the critical or non-official libertarian science. A similar process had occurred in Nazi Germany where the German dissident intellectuals suffered the repression of the state. In Brazil, the arrival of the military meant exile for many academics and intellectuals; this was the case for Josué de Castro and Milton Santos, just to mention two recognized, prestigious geographers who were affected by this situation.

Meanwhile, Antonio Núñez Jiménez became the official geographer of the Cuban regime[5], this presents positive and negative elements. On the one hand, the approach to power always complicates the independence of critical thinking. If

[5]As a public servant, he was the First General Secretary of the National Institute of the Land Reform from 1959 to 1961 and the General Secretary of the Academy of Sciences between 1961 and 1970.

we add to this that, for the Cuban case, the anti-castrist measures taken by the United States led to isolation and discredit, this went beyond the reach of an intellectual. In this sense, one can hypothesize that also scholarship remained limited in denouncing the responsibility of the United States in the poverty and/or dependence situation from one country to another.

Finally, a chronology of the major works that we quoted in this chapter shows how this scholarship did not develop in a vacuum. It was inserted in the contexts give the political conditions of the Latin American region and was likewise inserted in international circulation of knowledge, as it may be seen in Fig. 2.1.

2.3 Conclusions

As it has been observed, there was a type of critical discourse directly linked to geography, since the Forties in the Latin American region, which is little known and over which little has been reflected on and discussed within what now is considered as Latin American critical geography, as if critical discourse were anchored only in the English-speaking and French world of the Sixties.

We consider of great importance to recognize two early critical geographers such as Núñez-Jiménez and de Castro, who explicitly countered determinist, neo-Malthusian and colonialist geographies of the Forties and Fifties. We believe necessary that the new generations know these figures who contributed to the development of critical thinking in the Latin American region.

In Latin America, we consider it is pertinent and necessary to produce our own geographic tradition, while we question, criticize and debate based on the reading of original texts and not of common places discrediting all the works made in the past. Therefore, further interrogations should be developed on these scholars, such as: What was the impact that they had on their respective schools of geography? Did they know each other? Were they able to identify their points of coincidence or difference, or not? Was their concern a Latin American view of geography?

For that, we argue for the necessity of continuing the inquiry, over the generations that followed these two geographers in Latin America, and for the establishment of the Latin American scale as a horizon of development of these problematics. That is the reason why we ask ourselves questions that refer to the said scale, because it is well known that both authors counted with recognition in their respective countries, but what happens to that knowledge in the regional scale? Were they recovered or not? If it was not so, why were they unknown or forgotten? Why is there a feeling of rediscovering them for the first time? Or is it that we are barely rebuilding our Latin American tradition in geography?

All of this serves to investigate how the social processes raised as objects of study for geography beyond its traditional physical objects and descriptive concerns. Thus, it is necessary to reread the processes of the formation of the discipline, in order to overtake the simple empiricism and positivism which states that the data speak for themselves without any need for complex theory, as if the researcher did not come from a political and theoretical stance in his research, analysis and writing process of his research. Castro and Núñez presented outstanding examples of making geographies based on public engagement and non-banal theorization.

References

Bassols A (1957) XVIII Congreso Internacional de Geografía: Informe de la Delegación Mexicana. Revista Geográfica 47:60–74

Castree N et al (eds) (2010) The point is to change it. Geographies of hope and survival in an age of crisis. Wiley-Blackwell, Oxford

De Bomfim Albuquerque RP, de Mello Brito AD (2012) El XVIII Congreso Internacional de Geografía, Río de Janeiro, 1956. Actas del XII Coloquio Internacional de Geocrítica, pp 1–13

De Castro J (1957) La geopolítica del Hambre. Ensayo sobre los problemas alimentarios y demográficos de mundo. Solar Hachette, Buenos Aires

De Castro J (1984) A geografia da fome. Edições Antares, Rio de Janeiro

Dory D, Douzant-Rosenfeld D (1995) Geografía y geógrafos en Bolivia y Cuba: ensayo de sociología

histórica comparativa. De análisis Geográfico 27:57–73
Fermandois J (2016) Entre la geografía y el mundo: América Latina ante el sistema global. Estudios Internacionales 48(185):87–105
George P (1969) Sociología y Geografía. Editorial Península, Barcelona
Geoffrey JM (1980) The life and the thought of Isaiah Bowman. Archon Books, Hamden
Linhares Leite MY (1984) Apêndice à oitava edição. Uma contribuição da critica brasileira. A geografia da fome. Edições Antares, Rio de Janeiro
Macías FJ (2001) La enmienda Platt y la diplomacia española: crónica de una imposición neocolonialista a Cuba. Tebeto: Anuario del Archivo Histórico Insular de Fuenteventura 14:109–144
Moreira R (2007) Pensar e ser em Geografia. Ensaios de história, epistemologia e ontologia do espaço geográfico. Contexto, São Paulo
Núñez A (1960) La Geografía de Cuba. Editorial Lex, La Habana
Reclus E (1986) El hombre y la tierra. Fondo de Cultura Económica, México
Reclus E (1994) El hombre y la tierra. El pensamiento geográfico. Estudio interpretativo y antología de textos (de Humboldt a las tendencias radicales). Alianza Editorial, Madrid
Souza ML (2006) A Prisão e a Ágora: Reflexões em torno da Democratização do Planejamento e da Gestão das Cidades. Bertrand Brasil, Rio de Janeiro

After the Excitement of War: 'Disabled Veterans' in Modern Japan

3

Akio Onjo

Abstract

'Total war' has had many effects on human bodies and ideas as well as socio-spatial formation of nation-state. It generated countless wounded soldiers on the battlefield who sometimes became 'national heroes' in their homeland. However, after the war these war veterans with disabilities faced serious difficulties in their local communities. This chapter investigates some effects of the Russo-Japanese War on the formation of the nation-state in modern Japan by focusing on the experiences, consciousness, and acts of disabled veterans and the geographical imaginations of the people. Such an approach allows us to have insights into the social and political connectedness of nested scales from body to nation, through community and urban space.

Résumé

'La guerre totale' a eu des gros effets, non seulement sur les corps humains et les idées, mais aussi sur la formation socio-spatiale de l'État-nation. Elle a fait beaucoup de blessés de guerre sur les champs de bataille, qui sont devenu des 'héros nationaux' dans leurs pays. Pourtant, ils se sont trouvés en face de difficultés dans leurs communautés. Ce chapitre étudie l'influence de la guerre russo-japonaise sur la formation de l'État-nation au Japon modern, en explorant l'expérience, la conscience et les actes des blessés de guerre et les imaginations géographiques du peuple. Cette approche nous permet de saisir les connexions sociales et politiques d'échelles plurielles: le corps, la communauté, l'urbain et la nation.

Resumen

La 'guerra total' tuvo graves efectos, no solo sobre los cuerpos humanos y las ideas, pero también sobre la formación socioespacial del estado-nación. Esta hizo muchos heridos de guerra en los campos de batalla, que devinieran 'héroes nacionales' en sus países. Sin embargo, ellos encontraron dificultades en sus comunidades. Este capitulo estudia la influencia de la guerra rusojaponesa en la formación del estado nación en el Japón moderno, explorando la experiencia, la consciencia y los actos de los heridos de guerra y las imaginaciones geográficas del pueblo. Este abordaje nos permite de comprender las conexiones sociales y política de escalas plurales: el cuerpo, la comunidad, lo urbano y la nación.

A. Onjo (✉)
Faculty of Humanities, Kyushu University, Fukuoka, Japan
e-mail: aonjo@lit.kyushu-u.ac.jp

B. Schelhaas et al. (eds.), *Decolonising and Internationalising Geography*, Historical Geography and Geosciences, https://doi.org/10.1007/978-3-030-49516-9_3

Keywords

Disabled veterans · Russo-Japanese War · Homeland · Geopolitical social · Materiality · Discrimination · Impaired bodies · Public space

3.1 Introduction

'Total war' has had many serious effects on human practices and ideas in the modern world. These effects were felt not only by the soldiers on the battlefield but also by ordinary people in the homeland backing up the armed forces. Military bases have also influenced economic and social structures and cultural dynamics in a local and national context.

This paper examines how total war affected on the formation of the Japanese nation-state within its national territory. After the Meiji restoration, Japan waged two international wars: the Sino-Japanese War (1894–1895) and the Russo-Japanese War (1904–1905) which we discuss in this article. Japanese historians and social scientists regard this war as the first 'total war' or 'world war' in East Asia (Miyachi 1973; Iguchi 1998; Ôe 2001; Yamamuro 2005). According to them, it had also a strong influence on the formation of the Japanese nation-state, as well as on international relations in East Asia and worldwide. In the international relations Britain and France as the allied powers also involved in this war practically, and not only Asian people but also the people in Eastern Europe and Islamic areas were interested in the state of the war. Because this war was regarded as the struggle between 'white race' and 'yellow one'. So the manipulation of information became very important measure to gain the support of public opinion in the international scale. This war changed the character of war itself.

In the homeland a lot of Japanese people were pressed into service for the munitions plant and various local organizations raised their national identity. As the experiences of the Russo-Japanese War changed the disposition and historical-geographical imagination of the Japanese people, the central government had to launch a political and economic restructuring and plan the new social policy and work.

How can we consider the experiences of total war and how its impact on the everyday life of the people influenced the formation of nation-state at the beginning of the twentieth century? This paper drew on the notion of the 'geopolitical social' coined by Cowen and Smith (2009). They pointed out that modern geopolitics had treated two inter-relational issues: it was the discourses, practices, and acts of violence that helped not only to define European state territories and frontiers, and legitimate its territorial expansion (such as the idea of *Lebensraum* in Ratzel's *Politische Geographie*), but also to form the 'national society' within national territory. So 'geopolitical social' was the project of the making of national society as well as national territory. Ordering domestic social relations and social security is an essential element of the formation of the nation-state in European countries.

Cowen and Smith also emphasized that the geopolitical social originated in and through war. An example is the system of social insurance, including military pensions. This type of pension had expanded from soldiers to the whole population in daily life, in order to mobilize or unify the people who were divided by class, gender, or place specificity, to cultivate the allegiance of soldiers and ordinary people to the modern-state, and to control national sociality, solidarity, and security. Other 'state apparatuses' were also planned and started in the moment of war.

From the point of view of the geopolitical social, it is important to examine the complex and unfinished relations between war, citizenship, and the nation (Cowen and Gilbert 2008; Crooks et al. 2008). War became a crucial mechanism to get ordinary people to take part in various national projects and forge them into the 'good nation' who are approved for citizenship. As this socio-political process is situated in a particular socio-historical context and embedded in spatial scales from body to nation, it is contested, sometimes in contradictory ways. These scales are scaffoldings to the exercise of power relations and are always reconstructed by them. As Hastings and Thomas (2005) pointed out, constructing the good nation or citizen can privilege particular forms of embodied citizenship with a normalized body form (Woodward 1998).

War also generates countless wounded soldiers on the battlefield. As these wounded soldiers lacked a limb or arm, and lost their sight, they had to live as 'disabled people' in their everyday lives in the homeland. Their impaired bodies may not conform to the embodied citizenship, either ideologically or physically. Disability studies and geography have discussed how to treat the 'impaired body' in its social and environmental contexts (Imrie and Edwards 2007; Chouinard et al. 2010). In contrast with the stereotyped views of disability as a 'personal tragedy' and a condition requiring care, these studies emphasized and clarified the disabling social and environmental barriers in everyday space. However, understanding the materiality of the impaired body is indispensable when investigating the situation of disabled people in society and space, and their sentiments and disposition to the society (Hall 2000; Hansen and Philo 2007).

This paper focuses on the bodily experiences, acts, and emotions of disabled soldiers with the materiality of impaired bodies in their homeland in order to better understand the experiences of total war and the formation of national sociality in modern Japan. First, the geographical imaginations of the war by ordinary people and their dispositions and activities in wartime are traced. Second, complex and contested relationships between disabled soldiers and people in their 'homeland' are discussed. In particular, we consider the significance of the activity of peddling by disabled soldiers in the local context.

The research field is Sendai City, one of the main central cities in the Tohoku region in northeastern Japan. After the Meiji restoration, the central government installed one of the largest military bases in Sendai City in order to stabilize the local political and economic situation after the 'civil war'[1] and strengthen the defense of the northern frontier against the Russian Empire. Thus many military officers and soldiers and their families lived in Sendai City. There were also many conscripted soldiers and daily laborers in this city. The former were from farmers, artisans, and laborers, and the latter were the lower class and underclass who worked at logistic branch and in wartime called *Gunpu* on the battlefield. The socio-economic structures and the everyday life of people in Sendai City were greatly influenced by the trend of the army (Onjo 2015).

3.2 The Russo-Japanese War in the Local Arena: The 'Dream' or 'Reality' of Total War

How did Japanese people perceive the situations of this war in the 'homeland'? What kind of geographical imaginations did they produce from these extraordinary situations? Japanese people throughout the country became very excited by the news of the war. The newspapers played a crucial role in unifying all social strata beyond class or regional divisions and raised the nationalism of the people. Many of them reported only 'good' news from the battlefield. For example, though Shigetaka Shiga, who was a famous journalist, traveler, and geographer, went to the battlefield in mainland China and reported on military operations, he hardly reported on the actual cruel situations faced by the soldiers on the battlefield. Shiga made an effort to introduce a western modern geography to Japan and teach the geographical knowledge to the Japanese people. He thought that they should need this knowledge in order to preserve their national identity and survive keen international competitions under the circumstances of the world in the second half of the nineteenth century (Minamoto 1984; Nicolas-O and Nozawa 1993; Gavin 2001). Certainly he was not a simple-minded nationalist and expansionist, but it seems that

[1]Many feudal clans in the Tohoku region resisted and fought against the Meiji government.

even he could not look the reality of the battlefield in the face.

In many cities, affluent citizens launched various ceremonies in order to encourage or honor the soldiers and pray for or celebrate victory. Many citizens decorated their buildings and took part in parades carrying lanterns. These parades gradually became larger and more splendid. The citizens would hold these parades not only to encourage or honor the soldiers in the battlefield and celebrate battle victories, but also to enjoy themselves like 'doing the traditional local festival'. Sakurai (1997) pointed out that these parades were a kind of leisure activity for citizens in Tokyo City. As they had complaints about rising taxes under the wartime austerity policy, such parades were a manifestation of the people's energy and frustrations. They looked for an outlet for their excess energy. It sometimes became very powerful and dangerous to the public social order in urban spaces. Therefore, local governments started to regulate and control the unorthodox acts, such as drinking, wearing showy costumes, and fighting, of the people in the parade and ban these parades themselves.

In Sendai City too, neighborhood communities, secondary schools, companies, and trade associations organized the parades. At first, these were planned by each neighborhood community independently. It seemed that the parade and the acts of decorating buildings signified the rising of 'nationalism' and the allegiance to the nation-state among the citizens. However, neighborhood communities gradually began to compete in terms of the extent of their decorations and quarreled about how to do the parade. Therefore, the municipal government regarded these parades and other events as an obstacle or even harmful to unify various people in the 'homeland' where they had to back up the armed forces. The mayor of Sendai City, Tomohiro Hayakawa, asked the leaders of neighborhood communities not to make the parade more splendid and extravagant and cautioned the people:

> ...Many citizens stop their work and start to prepare a parade before the news of the victory. They waste a great deal of money and time. I wonder if this parade would be a kind of local festival and only a self-centered event by neighborhood communities. This festival goes against the irreplaceable devotion to the 'Japanese nation' during serious time of war. ... They must work hard according to their abilities and contribute to enhancing the national dignity. (*Kahoku Shinpo*, August 28, 1904)

Municipal government and the police helped to organize the joint parade of neighborhood communities to regulate and control the people's parade. Although a committee of the leaders of neighborhood communities was set up, they disagreed about the allocation of expenses, the order of the parade, and so on. While some neighborhood communities decided to cancel their participation in this parade, thousands of citizens participated, and the Green Gate was built near the station to receive the military units as the pride or the hero of the 'homeland' or 'Our Sendai City' (Fig. 1).

Although the mayor also asked these leaders to draw up a general arrangement and concrete measures for receiving the wounded soldiers, many of them did not respond to this request positively. Apparently, they saw the reception of the wounded soldiers as a burden and wanted to avoid this responsibility as the Japanese nation. While they projected an enthusiastic nationalism to the outside world during this imperialist war, they still did not become the 'loyal Japanese nation' in the domestic arena. So this nationalism of the people from below was situated in the social, cultural, and material contexts at the local scale, and the people sometimes behaved rashly, beyond the scope of the expectations of the ruling class. It did not always correspond to the nationality that the leading politicians and government officers thought of as the 'ideal Japanese'. The gulf between the two conceptions of nationality caused deep social antagonisms in postwar Japanese society.

When the central government concluded a peace treaty with Russia, the Japanese people had deep objections to its content and launched an anti-peace-treaty movement in the urban space. Many meetings featuring speeches against the anti-peace-treaty were held all over the country.

Fig. 3.1 Green Gate

In the case of Tokyo City, after such a meeting by the opposition, large-scale urban riots occurred in the streets and the public park, and police boxes were set on fire (Fujino 2015).

The government officers and the people had very different understandings about the result of this war. As the people were misled about the war, they believed that Japan defeated mighty Russia. The national consciousness of the people appears to have changed before and after this war. It was this war, which boosted the imperialistic consciousness and geographical imagination of the people as a 'first-rank nation state' or one of the 'eight great powers' in the world. Some government officers recognized this exaggerated enthusiastic nationalism from below as dangerous to the state.

Therefore, the central government had to manipulate the manifestations of these excess and amorphous energies of the people. In addition, as rapid urbanization changed the urban socio-spatial structures, urban hierarchy, and urban-rural relations as well as the life style or consumption habit among urban residents, Japanese society thus faced with a lot of urban social and political problems (Ohishi and Kanazawa 2003; Onjo 2014).

In this situation, one of the urgent postwar social problems was how to treat the sick and wounded veterans who returned to their 'homeland'. While the about 85,000 soldiers were killed, the number of wounded was roughly 150,000. Wounded veterans could not earn their daily wages or food due to their impaired bodies and survived only on military pensions or temporary rewards. Thus, many of them were forced into poverty. The poverty of disabled veterans and their families was a serious social problem, and the social antagonism became intense in urban and rural areas alike. Some disabled veterans got into violent trouble in daily life because they felt abandoned by the government.

If the central government had been to unable to meet the needs of the disabled veterans, the veterans might have abandoned their allegiance to the Meiji emperor and the 'Great Japanese

Empire'. Government officers also seemed to regard these disabled bodies as a 'dangerous entity', 'an object of fear', and even an obstacle to constructing the nation as embodied citizenship (Hastings and Thomas 2005).

So they made a program to re-integrate these disabled veterans into the state apparatus. They researched the living conditions of the disabled veterans in each prefecture, reconsidered the system of military pensions, and adopted some social policies on them. At the national Diet, on April 1906, the central government passed a law on 'the Hospital for Disabled Veterans (*Haiheiin* 廃兵院)'. One of the most important statesmen and the general of the army, Aritomo Yamagata, was willing to organize this hospital on the model of Les Invalides, the disabled veterans' hospital in Paris established by Louis XIV in 1676.

The ministry of the army planned to build this type of hospital in three places (Tokyo, Osaka, and Kokura) to accommodate 300 disabled veterans. In the end, however, only one was established in Tokyo in 1907 (Yamada 2005). The central government expected many to apply to enter *Haiheiin*, but the number of the applicants did not even reach the number of space available. One reason was that their families could not take care of them at *Haiheiin*. It was particularly inconvenient for those who did not live in Tokyo. They believed that entering *Haiheiin* might mean being separated them from their families, war comrades, and local communities.

3.3 There Is no Such Thing as '*Homeland* (郷土)' for Disabled Veterans

It was estimated that there were about fifty wounded veterans in Sendai City (*Tohoku Shinbun*, October 15, 1906). As they had lost their arms or legs or had been blinded, they became 'disabled veterans' in their homeland. Many of them were lower ranking soldiers. Certain veterans were rewarded or aided by the central government, but the sum of money they received was not enough to live on. As they could not return to their original jobs, were out of work, and could not find new jobs due to their disabled bodies, their families had trouble making a living. As career officers were eligible for a military pension, their circumstances differed from those of lower ranking disabled veterans. Although both of them had fought against Russian soldiers on the same battlefield, the sharp contrast between the living conditions of each was seen as a desperate situation in postwar society.

Most of them spent all day alone at home and avoided contact with others, because the everyday living space became a 'disabling space' for their impaired bodies in the city. As other citizens gawked at their impaired bodies, the feelings of shame and embarrassment over their bodies would cause them to withdraw from public spaces (Hawkesworth 2001). They suffered from loneliness and alienation from 'normal society' as well as physical pain and tough economic conditions. They needed spiritual well-being and social interaction, not to mention physical care. A disabled veteran wrote a letter to a newspaper as follows.

> I was injured on the battlefield and was then taken a captive. …I lost my left hand and leg in the enemy hospital…. When my father saw my impaired body, he lost his mind. I can't earn our daily bread due to this impaired body. Now I am in charge of helping my wife. I still haven't received a reward from the government. (*Tohoku Shinbun*, December 23, 1906)

Newspaper reporters insisted that 'local society' had a duty to mitigate and improve this unpleasant situation, in addition to central and local governments. The miserable situations of disabled veterans and their families were regarded as social and humanitarian problems in Sendai City. Local newspapers frequently reported the disabled veterans' difficult economic conditions and miserable everyday life, and made an appeal to the sympathy, benevolence and solidarity of the career officers or the citizens to assist them. A disabled veteran told to a female newspaper reporter:

> …I am very lucky to be employed at the municipal office. … Please consider taking action so that other disabled veterans can make a living. (*Kahoku Shinpo*, July 3, 1906)

The governor of Miyagi prefecture's wife, who was the head of 'the Association of Women' Patriots (愛国婦人会)' in Miyagi prefecture, read these articles and recognized that the difficult situation of disabled veterans was a social problem that needed to be tackled. The members of this association went to the houses to see the disabled veterans, but had no means or funds to assist them.

As many researchers have already pointed out, women took part in the various 'public spheres' in wartime, either voluntarily or involuntarily. The Japanese government asked upperclass women to become 'patriots' or 'mothers of the nation' on a local scale. They organized parades, visited the houses of the soldier's families, and collected contributions from the citizens to assist the families whose husbands were conscripted. However, these women and the soldier's families seemed to have a delicate relationship, because many lower-class soldiers' families sometimes felt repelled toward the attitudes and dress of upperclass women. Although lower-class people did not seem to have a definite 'class identity' at the time, they felt a clear distinction between 'us' and 'them'. While the war enhanced nationalism and strengthened the sense of solidarity among the people, these social ties at the local scale might not have been so strong. These assisting activities did not necessarily spread to the citizens. Many of them were less and less interested in and even forgot about the miserable condition of disabled veterans and their families after the excitement of wartime.

While central and local governments tried to integrate disabled veterans in neighborhood communities and urban society, it was a very difficult task under postwar conditions. A disabled veteran told a reporter in anger:

> …In general, Japanese people are apt to become impassioned and then to cool down very quickly. During the war, Japanese people had the great enthusiasm or excitement about the war, lamented the respected glorious deaths of soldiers in battle, and praised the honorable wounded soldiers as heroes. However, when they returned to peaceful everyday life, their attitudes changed dramatically and their enthusiasm cooled immediately…. I'm afraid that many people regarded these wounded veterans as no longer 'honorable wounded heroes', but only 'disabled, handicapped or disfigured people'. (*Kahoku Shinpo*, June 26, 1906)

The abrupt change in the citizens' attitude was a humiliating and devastating experience for disabled veterans. They felt that alienation from the homeland and the indifference from or discrimination by their fellow citizens was even worse than the difficulty of living in impaired bodies and tough economic conditions.

How did they confront both the deficient state policy and the citizens' indifference? In December 1906, a disabled veteran planned to establish the '*Sendai Merchant House by the Disabled Veterans* (*Sendai Haihei Shokan* 仙台廃兵商館, or *Sendai Haiheikan*)' in order to assist disabled veterans living in and around Sendai City. The name seems to have been borrowed from the *Haiheiin* managed by central government, but its aims and activities were very different from those of *Haiheiin*. He said the following:

> The disabled veterans who belonged to the rank and file in the army become the poor in their homeland. Then it occurred to me to establish the *Haiheikan*. We shall start two businesses: First, we have applied to the army to sell leftover rice. Luckily, we will receive the contract with the army starting next year. Second, we peddle the daily necessities or cosmetics provided by local shops. We shall accommodate four disabled veterans and buy two large carts and make a uniform for peddlers. The carts and uniform are painted in the color of khaki, evoking the army. One peddler, who lost his arm, pulls the cart, and the other, who lost his leg, takes the orders…. The incomes are divided equally among the members. This money may be the funds for economic independence in the near future. (*Tohoku Shinbun*, November 28, 1906)

This plan aimed to develop the means of production and reproduction that enabled disabled veterans survive and empower themselves in everyday life. They also rented a house near the *Haiheikan* to live together and economize on the cost of living. They might have been able to enhance the mutual solidarity that was lost in the homeland. At first, eight disabled veterans peddled the small goods at girls' high schools and military officers' houses, earning between six and twelve yen in a day. Leftover rice was a basic food for the underclass and the labor class in urban society in those days. Many of them preferred leftover rice to the imported rice from China and East Asia, because of the difference in taste and a need to economize on cooking costs (Onjo 2014). There were five to seven traders who dealt with leftover rice in Sendai City and in 1910, the *Haiheikan* had approximately 25 regular customers per day (*Kahoku Shinpo*, February 23, 1910). Later in order to expand its business, it built a facility for raising pigs and cattle. After 5 years, this small merchant house developed to be one of the second-ranking food shops in Sendai City (Onjo 2015).

Why did a disabled veteran launch this project? Was the reason only to do with economic and living conditions? When this merchant house was started, newspapers urged the citizens and other shop owners to aid the *Haiheikan*'s activities emotionally and materially, but the founder said:

> Everyday acts of peddling are a very hard job for disabled veterans for two reasons. One is that they don't have any experience with commerce and sometimes have misunderstandings with their customers. … The other is that a bad muddy and gravelly road is an obstacle for their impaired or clumsy bodies and they have to go around this road. … After the establishment of *Haihei shokan*, some citizens gave us some money and materials. But we, our wounded veterans, made an effort to work and manage this business by ourselves. We have no intention of being supported through material aid or donations by the municipal office and the public.… When we peddle, we attach an importance to sincerity.… We have no intention of selling any luxurious goods at girls' schools. (*Tohoku Shinbun*, December 13, 1906)

He emphasized the importance of working continuous with use of the impaired bodies and the mutual aid:

> … We shall work our impaired bodies effectively from now on for all our life.… If this project arrives at its basic aim, we shall use the earnings to pursue the mutual aid, including the disabled veterans who could not work at all because their bodies were the most seriously impaired. (*Kahoku Shinpo*, December 1, 1906)

He thought that disabled veterans need not only achieve economic independence, but also to take great care with their appearance. As he painted their carts in khaki and made a uniform, he might have been possessed by his pride in these men as veterans of the Japanese imperial army. However, the passion that pushed him on this project might be have been rooted in a profound despair and a strong anger at the Japanese government policy on disabled veterans and the discriminative attitudes or gazes of the people in the local community or 'homeland'. By engaging in peddling day after day, they could always move around the street and public space, expose their 'impaired bodies' to the public gaze, and establish and maintain face-to-face relations with other citizens.

'Ordinary citizens' who had the special consciousness of being the 'first-class nation-state' in the world wanted to avoid seeing these impaired bodies in the daily 'normal' urban landscape. The reason was that seeing the 'impaired bodies' of disabled veterans might have awoken them to realize the miserable reality and memories of the war, which did not suit their geographical imaginations of 'glorious newly imperial Japan'. Impaired bodies were excluded from public space, because they disturbed the 'dream' of the people.

While many people indulged themselves in the dream of the Japanese Empire as a great power, disabled veterans had to live the reality of impaired bodies in postwar period. Disabled veterans had to resist the barriers and discrimination imposed on them by 'normal' society, which wanted to erase their existence. The everyday acts of peddling seem to have been a kind of 'soft weapon' by which to make disabled veterans visible. The social practices of peddling and its spatial mobility might have helped to recover their bodily practices or performance by training impaired bodies. Additionally, it could enhance the self-esteem of disabled veterans, as well as the social recognition of them. It could build new social interactions among the veterans who were isolated from each other, and between veterans and the citizens. The acts of peddling and other commerce made them visible and provided a potential means by which disabled veterans could overcome the barrier of social indifference and taken-for-granted social gazes. The peddling empowered disabled veterans to 'erase their own erasure'.[2]

By contrast, in the *Haiheiin*, while the disabled veterans could be nursed with great care, they might be separated from their local communities, confined to a closed space, and rapidly forgotten by the nation.

In their 'homeland', many wounded veterans were treated as disabled, not as local heroes. In real life, there was no such thing as 'homeland' for disabled veterans. The trope and narrative of 'homeland' means a local community where people can have strong emotional attachments, invoke common bloodlines and homogeneity, and make intimate social ties, as well as being a birthplace. However, it is not always natural, but rather invented ideologically. The circumstances of the disabled veterans exposed the 'fictitious nature' of the notion of homeland. These contradictions between 'dream' and 'reality' led to the plan for new social policy and work within the national territory.

After the Russo-Japanese War, some bureaucrats of the Ministry of Domestic Affairs realized again the important functions of rural and urban local communities in the formation of the nation-state (Miyachi 1973; Sakurai 1997). They believed that the ideological apparatus of the 'homeland' was one of the basic scaffoldings for integrating various people into the nation-state and constructing the Japanese nation. However, it had to be reformed socially and economically as well as ideologically. Therefore, they worked to intervene in and improve the habits, customs, social relations, and mentality of ordinary people at the neighborhood scale in rural and urban world. For example, a young men's association was organized and took the place of the traditional groups. This movement from above by bureaucracy was called 'The Reform Movement of Local Communities (*Chiho Kairyo Undo*)'.

Since then other government projects started to focus on the reproduction of labor power. The plan was to build the public retail market and social housing in order to provide fresh foods, regulate prices by an official rate, and erase the traditional trade customs. This policy tried to transform ordinary people into 'modern and rational consumers'. Government officers and some politicians began to regard the everyday life of the people as an essential subject of their national policy (Harada 1997).

[2]Smith 1992: 58. Smith discussed the invention of a 'homeless vehicle' that could make homeless people visible by enhancing their spatial mobility.

3.4 Concluding Remarks

In general, total war restructures the consciousness and practices embodied in the people, as well as the existing social, economic, and political systems. Its effects range widely from the body to global scale. While it enhances 'nationalism' and national identity beyond the various differences ideologically, it devastates innumerable human bodies physically and leads to social antagonism and deep discrimination or exclusion within national territory. It invents the new state system and national sociality within the territory, as well as the remaking of inter-state relations in the international context. Therefore, the restructuring processes that created the modern world system are not straightforward, but contested and contradictory.

This paper has examined some of the deeper effects of the Russo-Japanese War on the formation of the nation-state in modern Japan by focusing on the experiences, consciousness, and acts of disabled soldiers and the geographical imaginations of the people. We must further investigate the various aspects of the acts of violence within national territory, as well as between nation-states, in order to understand the complicated relations between war, state, and the nation in modern world.

References

Chouinard V, Hall E, Wilton R (eds) (2010) Towards enabling geographies. Ashgate, Farnham and Burlington

Cowen D, Gilbert E (eds) (2008) War, citizenship, territory. Routledge, New York

Cowen D, Smith N (2009) After geopolitics? From the geopolitical social to geoeconomics. Antipode 41 (1):22–48

Crooks V, Dorn M, Wilton R (2008) Emerging scholarship in the geographies of disability. Health & Place 14(4):883–888

Fujino Y (2015) Toshi to Bōdō no Minshushi Tokyo 1905–1923 [Popular history of the city and the Riots Tokyo 1905–1923]. Yushisha, Tokyo [Jap]

Gavin M (2001) Shiga Shigetaka 1863–1927. The Forgotten Enlightener. Curzon, Richmond

Hall E (2000) 'Blood, brain and bones': taking the body seriously in the geography of health and impairment. Area 32(1):21–29

Hansen N, Philo C (2007) The normality of doing things differently: bodies, spaces and disability geography. Tijdschrift voor Economische en Social Geografie 98 (4):493–506

Harada K (1997) Nihon kindai Toshishi Kenkyu [Historical study on modern Japanese city]. Shibunkaku, Kyoto [Jap]

Hastings J, Thomas H (2005) Accessing the nation: disability, political inclusion and built form. Urban Stud 42(3):527–544

Hawkesworth M (2001) Disabling spatialities and the regulation of a visible secret. Urban Stud 38(2):299–318

Iguchi K (1998) Nichirosensō no Jidai [The era of the Russo-Japanese War]. Yoshikawakōbunkan, Tokyo [Jap]

Imrie R, Edwards C (2007) The geographies of disability: Reflections on the development of a sub-discipline. Geogr Compass 1(3):623–640

Minamoto S (1984) Shigetaka Shiga 1863–1927. In: Freeman T (ed) Geographers biobibliographical studies, vol 8. Mansell Publishing, London, pp 95–105

Miyachi M (1973) Nichiro Sengo Seijishi no Kenkyu [Study on political history after the Russo-Japanese War]. Tokyo Daigaku Shuppankai, Tokyo [Jap]

Nicolas-O G, Nozawa H (1993) Shigetaka Shiga. Eratosthène, Lausanne

Ôe S (2001) Sekaishi toshiteno Nichirosensō [The Russo-Japanese War as world history]. Rippushobō, Tokyo [Jap]

Ohishi K, Kanazawa F (ed) (2003) Kindai Nihon Toshishikenkyu [Historical studies on modern Japanese cities]. Nihonkeizaihyoronsha, Tokyo [Jap]

Onjo A (2014) Eating rice and feeding the city: 'body politics' in modern Japan. In: Shimazu T (ed) Languages, materiality, and the construction of geographical modernities. Department of Geography Wakayama University, Wakayama, pp 29–40

Onjo A (2015) Nichiro Senji Sengo no Sendai [Sendai city before and after the Russo-Japanese War]. Kukan, Shyakai, Chirishiso [Space Soc Geogr Thought] 18:17–28 [Jap]

Sakurai R (1997) Taisho Seijishi no Syupatsu. [Start of Taisho political history]. Yamakawashuppan, Tokyo [Jap]

Smith N (1992) Contours of a spatialized politics: homeless vehicles and the production of geographical scale. Soc Text 33:54–81

Woodward R (1998) 'It's a man's life!': soldiers, masculinity and the countryside. Gend Place Cult 5:277–300

Yamada A (2005) Nichirosengo Meijikiniokeru Haihei no Seikatsu Mondai to Haiheiinn Seisaku no Tokushitsu [Political characteristics of disabled veterans after the Russo-Japanese War in the Meiji era]. Nihon Fukushi Senmon Gakko Kenkyu Kiyo 13:65–82 [Jap]

Yamamuro S (2005) Nichirosensō no Seiki [The century of the Russo-Japanese War]. Iwanamishoten, Tokyo [Jap]

Newspapers

Kahoku Shinpo. (microfilm) Miyagi Prefecture Library

Tohoku Shinbun. (microfilm) Miyagi Prefecture Library

Indian Ocean Small Islands Along the Postcolonial Trajectory: Chagos and the Maldives

4

Marcella Schmidt di Friedberg and
Stefano Malatesta

Abstract

In 2009, Robert Kaplan defined the Indian Ocean Region (IOR) as "the central internship for the challenges of the twenty-first century". For Bouchard and Crumplin (2010), the issues now playing out in the Indian Ocean mean that it can be "neglected no longer" within world geopolitics. The crucial historic transition underway in the area is being driven by the network of relations among the "big players" on the regional chessboard: the USA, China, and India. Both the geographical nodes at which these relations unfold and new narratives shaping the geographical imaginary of the region are key to understanding these contemporary historical processes. In this essay, we examine the main historic and geographic processes that laid the ground for the current period of transition: the renewed centrality of the Indian Ocean, Sino-Indian rivalry (Brewster 2015) and the geopolitics of environmental crisis. We focus here on two archipelagos (the Chagos Islands and the Maldives) that have been described as "unsinkable aircraft carriers" within this scenario. In recent decades, Chagos and the Maldives have become key nodes in the new geopolitical cartography. In 1971, the US Navy built a military base on Diego Garcia and, since the beginning of the "Indian Ocean Cold War", the Maldives have been situated at the fringe between Indian military supremacy and China's influence on the economies of peripheral counties. This focus allows us to investigate contemporary transitions and to propose an alternative reading of the conflict for hegemony currently underway between the "big global players".

Résumé:

Les îles de l'océan Indien le long de la trajectoire postcoloniale: Chagos et les Maldives

En 2009, Robert Kaplan a défini la région de l'océan Indien (RIO) comme «le stage central pour relever les défis du XXIe siècle». Pour Bouchard et Crumplin (2010), les enjeux qui se posent aujourd'hui dans l'océan Indien font qu'il ne peut plus être «négligé» dans la géopolitique mondiale. La transition historique cruciale en cours dans la région est portée par le réseau de relations entre les «grands acteurs» de l'échiquier régional: les États-Unis, la Chine et l'Inde. Les nœuds géographiques où se déroulent ces relations et

M. Schmidt di Friedberg · S. Malatesta (✉)
"Riccardo Massa" Department of Human Sciences and Education/Marine Research and High Education Center (MaRHE), University of Milano-Bicocca, Milan, Italy
e-mail: marcella.schmidt@unimib.it

M. Schmidt di Friedberg
e-mail: stefano.malatesta@unimib.it

B. Schelhaas et al. (eds.), *Decolonising and Internationalising Geography*, Historical Geography and Geosciences, https://doi.org/10.1007/978-3-030-49516-9_4

les nouveaux récits qui façonnent l'imaginaire géographique de la région sont tous deux essentiels à la compréhension de ces processus historiques contemporains. Dans cet essai, nous examinons les principaux processus historiques et géographiques qui ont jeté les bases de la période de transition actuelle: la centralité renouvelée de l'océan Indien, la rivalité sino-indienne (Brewster 2015) et la géopolitique de la crise environnementale. Nous nous concentrons ici sur deux archipels (les îles Chagos et les Maldives) qui ont été décrits comme des "porte-avions insubmersibles" dans ce scénario. Au cours des dernières décennies, les Chagos et les Maldives sont devenus des nœuds clés de la nouvelle cartographie géopolitique. En 1971, l'US Navy a construit une base militaire sur Diego Garcia et, depuis le début de la «guerre froide dans l'océan Indien», les Maldives sont situées à la frontière entre la suprématie militaire indienne et l'influence de la Chine sur les économies des régions périphériques. Cette réflexion permet d'explorer les transitions contemporaines et de proposer une lecture alternative du conflit d'hégémonie en cours entre les «grands acteurs mondiaux».

Resumen:

Las islas del Océano Índico a lo largo de la trayectoria postcolonial: Chagos y las Maldivas

En 2009, Robert Kaplan definió la Región del Océano Índico (IOR) como "the central internship for the challenges of the twenty-first century". Para Bouchard y Crumplin (2010), los problemas que se plantean actualmente en el Océano ya no se pueden "descuidar" en la geopolítica mundial. La crucial transición histórica en curso en la zona está siendo impulsada por la red de relaciones entre los "grandes jugadores" del tablero de ajedrez regional: Estados Unidos, China e India. Tanto los nodos geográficos en los que se desarrollan estas relaciones como las nuevas narrativas que conforman el imaginario geográfico de la región son claves para entender estos procesos históricos contemporáneos. En este capitulo se examinan los principales procesos históricos y geográficos que sentaron las bases del actual período de transición: la renovada centralidad del Océano Índico, la rivalidad sino-india (Brewster 2015) y la geopolítica de la crisis ambiental. Nos centramos aquí en dos archipiélagos (las Islas Chagos y las Maldivas) que han sido descritos como "portaaviones insumergibles" dentro de este escenario. En las últimas décadas, Chagos y las Maldivas se han convertido en nodos clave de la nueva cartografía geopolítica. En 1971, la Armada estadounidense construyó una base militar en Diego García y, desde el comienzo de la "Guerra Fría del Océano Índico", las Maldivas han estado situadas al margen entre la supremacía militar india y la influencia de China en las economías de los condados periféricos. Este enfoque nos permite investigar las transiciones contemporáneas y proponer una lectura alternativa del conflicto en curso entre los "grandes actores globales".

Keywords

Indian ocean · Small islands · Chagos · Maldives · Geopolitics · Post-Colonial chessboard · China · India

4.1 Telling a Story: Small Islands and "Big Players"

On the evening of 30 August 2018, the channel dividing Kaafu Atoll (in the Maldives) hosted the inauguration of Sinamalé, a two-km-long road bridge connecting two islands: the capital Malé and Hulhumalé. At the end of the 1990s, the Maldivian government designated Hulhumalé as a key area for urban and retail development, foreign investment, and international tourism. A spectacular Chinese fireworks display opened the inauguration ceremony. The bridge is a mega-infrastructure (in relation to the average size of the Maldivian islands), an iconic monument, and a symbol of the "everlasting friendship" linking a Small Island State (the Maldives) with one of the "big players" in the Indian Ocean (China): The political role of infrastructures and the equilibrium between big and small players are the topics we set out to address in this paper. We bring a geo-historical perspective to bear on our reading of the "geopolitical changes" (Mohan 2010) that have taken place in the wake of the de-colonization process, across this "central stage" of our contemporary era (Kaplan 2009, 2010).

In the first part of the chapter (paragraph 2 and 3), we go beyond a Mahanian perspective to outline the historical processes and geopolitical tensions that shaped the geography of maritime spaces in the Indian Ocean after the formal ending of British imperial dominance: a geography structured around the so-called "triangular condominium" (India, China, and USA), the system of power relations that still controls the region today.[1] In the second part (paragraph 4 and 5), we discuss the strategic relevance of archipelagic regions and the main narratives defining the Indian Ocean as a "large-scale ocean basin-centric region" (Bouchard and Crumplin 2010: 26). Our analysis of the regional framework draws together contributions from the fields of geography, social anthropology, regional studies, history, and international relations. We also rely on the interpretative framework of *critical geopolitics* (Ó Tuathail 1996), which offers a reading of the key narratives underpinning varied and distinctive discourses about space and politics (Agnew 1998; Ó Tuathail 2005).

4.2 A Few Decades of Post-colonial History

The second half of the 1960s was a crucial period in the contemporary history of the Indian Ocean (Pearson 2003; Alpers 2014): British Imperial hegemony gave way, de facto, to the quest for a new geopolitical equilibrium. The historical processes implicated in the transition to the post-colonial[2] regime were largely shaped by this struggle to establish a new system of power relations among the major regional and supra-regional actors. Experts from the Task Force on Indian Ocean Security have identified the "emergence of a new security agenda" (Australia India Institute 2013) as the main geopolitical driver guiding the historical phase from the end of the Cold War to the beginning of the twenty-first century, with a climax occurring in the late 1960s.

In 1967, the British troops left Aden, and 1968 saw the beginning of the British armed forces' withdrawal from the major military bases East of Suez. This phase conventionally labelled "British disengagement" from the Indian Ocean, gave rise to a complex sequence of post-colonial dynamics. In keeping with our aims in this paper, we go about reading this complexity by focusing on the stories of two archipelagos (Chagos and the Maldives), which initially followed two different historical trajectories. As we discuss in the second part of the paper, in the twentieth-first century have come to share the same destiny: together, they rank among the hotspots of the regional security system, fulfilling, similarly to other archipelagic regions such as the Andaman groups, a strategic function within the contemporary geo-strategic scenario (Fig. 4.1).

Chagos is a chain of about sixty coral atolls lying between the latitudes of 4 °S and 7 °S and the longitudes of 7 °E and 72 °E. Along with Mauritius, the archipelago was part of the British Empire from 1814 until the 1960s. In 1965, it was included in the newly established British Indian Ocean Territory (BIOT), an overseas territory that grouped together a few former British colonies in the region. Since the pre-colonial period, the only stable human settlements have been located on the island of Diego Garcia. Chagos' colonial and post-colonial trajectories

[1] Saudi Arabia, Indonesia, and Australia are the other big players within this power structure. However, we focus here on India, China, and the USA, because the "triangular con-dominium" is a key geographical factor in the tendency to understate the geopolitical role of small islands.

[2] By using the term "post-colonial" we mainly refer to the historical phase after the British Empire "disengagement" from the regional chessboard.

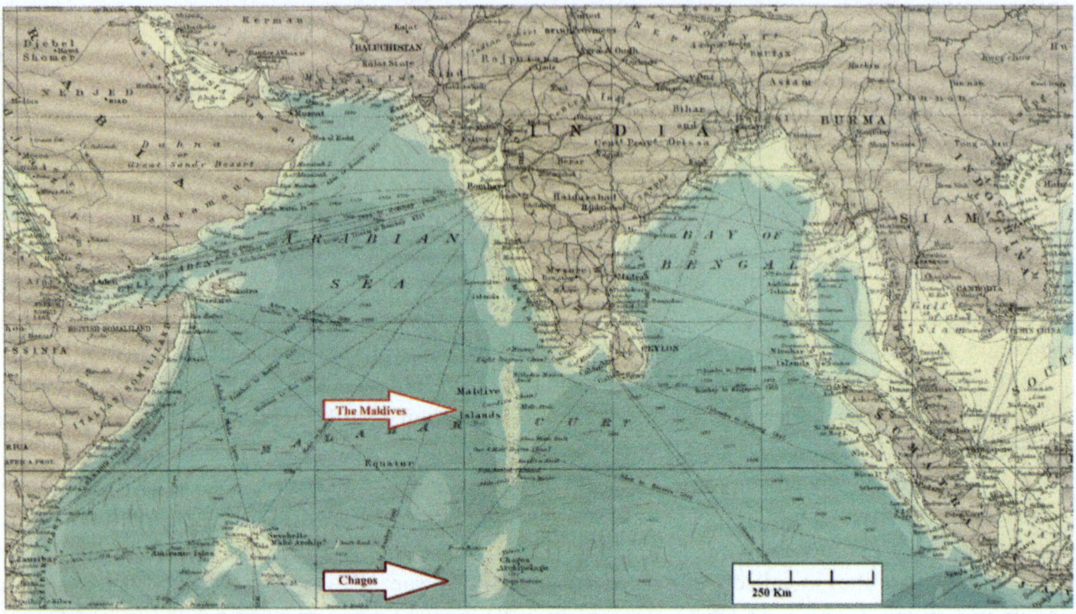

Fig. 4.1 Original title "Indian Ocean on Mercator's projection". The map stresses the strategic positions of the two archipelagos within the regional scenario during the British Empire, by showing ocean steamer routes with distances or submarine cables *Source* The Edinburgh Geographical Institute, John Bartholomew & Co. *The Times atlas*, London, 1922. *Credits* David Rumsey Map Collection, www.davidrumsey.com

are closely bound up with those of Mauritius. According to Jeffery, "during independence negotiations with Mauritian politicians in 1965, the UK government persuaded Mauritian politicians to agree to the excision of the Chagos Archipelago from Mauritius […] Retaining the Chagos Archipelago as part of the new British Indian Ocean Territory (BIOT), the UK government then signed an agreement with the US government to make the largest Chagos island, Diego Garcia, available for a major US military base" (Jeffery 2007: 954). Following this agreement, the archipelago became a key element of US military control over the region, reinforcing the legitimization of the USA as the new guarantors of maritime security in the area: "Diego Garcia was completely depopulated by 1971 to make way for the construction of the US military base, which has subsequently become central to the US-led so-called 'War on Terror' and has been used extensively in military attacks on Afghanistan since 2001 and Iraq since 2003" (*ibidem*). Since the 1970s, groups of exiled Chagossians had been engaged in a legal battle for financial compensation and the right to be resettled on Diego Garcia. However, an extensive review of this history would lead us away from our main topic.[3] Numerous studies (Frost

[3] We briefly summarize the key stages in this legal case. In 2003–4, Chagossians failed to win economic compensation. In "2000 and 2006, judicial reviews successfully challenged the immigration legislation preventing non-authorized persons from entering the territory, but the UK government appealed to the House of Lords, which reinstated the immigration legislation in 2008" (Jeffery 2013: 303): In 2004, the UK Government prevented the Chagossians from going back to their island, by invoking the "Royal Prerogative", which is still in force in former colonial territories. In 2010, the UK government established a Marine Protected Area (MPA) around the archipelago, which would justify any future opposition to Chagossian resettlement, in the name of biodiversity and environmental protection. In 2012, the European Court of Human Rights ruled that it had no jurisdiction to re-examine the case. (Frost and Murray 2015; Jeffery 2007, 2010, 2013): Following a UN general assembly resolution, in September 2018 the territorial dispute between Mauritius and the UK came before judges at the International Court of Justice (ICJ) in The Hague. The case is still open, and UN resolution can be interpreted as a sign of Britain's decreasing international influence in the area, after the Brexit vote.

and Murray 2015; Jeffery 2007, 2010, 2013) have documented this long and troubled case, which is still ongoing, with a new episode playing out at the time of writing.[4]

The Maldives is a group of twenty-six coral atolls (made up of approximately 1190 islands), lying between the latitudes of 1 °S and 8 °N, and the longitudes of 72° and 74 °E. The British Empire took control of the archipelago in the late eighteenth century, and in 1887 it officially became a British Protectorate. This juridical status lasted until 1953, when Muhammad Amin Didi was proclaimed president of the "First Republic". The pioneer republican period lasted until 1954, followed by the restoration of the Sultanate and, consequently, of British control. The Sultanate officially ended in 1965, and in 1968 the first republican parliament was elected. Since 1968, the country has been run by a succession of six presidents: Ibrahim Nasir (1968–1978), Maumoon Abdul Gayoom (1978–2008), Mohamed Nasheed (2008–2012), Mohammed Waheed Hassan (2012–2013), Abdulla Yameen (2013–2018), and the current, President Ibrahim Mohamed Solih. The period between the "First Republic" and the end of the Protectorate is crucial to understanding the archipelago's postcolonial geopolitical role. In fact, in 1957, the Royal Air Force obtained permission to restore a military air base on the island of Gan, in the atoll of Seenu, the southernmost of the archipelago and the closest, geographically speaking, to Chagos. The base remained operative until the withdrawal of British troops (in 1976) in line with the strategy of "disengagement" initiated about ten years earlier.

4.3 The Indian Ocean Region: A Central "Oceanic" Stage

Robert Kaplan (2009, 2010), in a discussion of the rivalry among regional and global big players, defined the Indian Ocean Region (IOR) as the central internship for the challenges of the twentieth century. He saw these challenges as mainly concerning the geostrategic domain, while in the second part of this paper we ourselves set out to review a broader spectrum of geographical challenges. Nonetheless, Kaplan's analysis raises a number of key issues, including the importance of studying Indian Ocean as a region, the differences between big players and weaker states, and the dominant refrain of maritime security in the most powerful narratives shaping the regional chessboard.

According to Bouchard and Crumplin (2010), the Indian Ocean is a maritime region bounded by continental Africa, continental Asia, the Indonesian Archipelago and Australia, and the 60th parallel south, thus bordering on the area regulated by the "Antarctic Treaty". Rumley, Doyle and Chaturvedi proposed alternative definitions for the region based on the centrality of oceanic spaces, using either a formal criterion to include a group of states bordering on the Indian Ocean, or a functional one, comprising all "states that have an interest in the maximisation of the Ocean's security" (Rumley et al. 2012: 4). In both cases, ocean space is viewed as the central object in the construction of the geographical region. These authors sketched out a "maritime regionalism paradigm" which envisages among other elements, that: 1. the ocean is the main spatial feature; 2. maritime security is a political, and multidimensional, priority; 3. the relations, rivalries, and cooperation among the players define the IOR as a space of international interactions. Beyond formal and functional classifications, as argued in *critical geopolitics*, regionalization is above all a process of "geo-writing" (Ó Tuathail 1996; Agnew 1998, 2000): Thus, the narratives shaping "geo-writing" in the IOR include: obsession with maritime security; power relations among states; and the centrality of oceanic spaces.

Bouchard and Crumplin argued that the IOR has recently regained prominence within the international debate, defining it as "large-scale ocean basin-region". They also pointed out the tendency to neglect the IOR in the "academic literature on geopolitics, international relations and geostrategy, as well as in the public news arena" (Bouchard and Crumplin 2010: 28) during

[4]To follow the most recent developments, see https://www.icj-cij.org/en.

the post-colonial period. The current scenario has pushed the IOR to "the forefront of world geopolitics […] as an area of crucial geostrategic importance [that will] remain so at least for many decades to come" (*ibid*: 27).

In the view of Bouchard and Crumplin, experts and analysts should focus on four drivers underlying this newly perceived centrality of the Indian Ocean: the rise of India as a regional power; China's oceanic strategies; military engagement (and dis-engagement) by the USA; and the security of maritime lines of communication and chokepoints (such as the Ormuz and Malaccan straits). Hence, the key areas of analysis must include regional equilibrium, the three big players, a dominant narrative (maritime security), and, above all, two geographical features defining the IOR as a space of communication and transport: maritime lines and strategic nodes. In this section, we discuss the tension and equilibrium within the triangular con-dominium (India, China, and the USA), while, in the following sections, we argue that archipelagic regions represent crucial nodes within this security system.

The affirmation of India as a regional power relies on its framing of the ocean as an "Indian" space and the role of Indian naval forces in keeping it safe (Kaplan 2009; Mohan 2010; Brewster 2015): The expansion of India's "naval presence" (Kaplan 2009) across the IOR has been achieved via networking with other countries, bilateral alliances, and strategic investments. In 2014, a military exercise called *Milan 2014* saw the involvement of navies and coastguards from thirteen countries across the IOR (Bangladesh, Sri Lanka, the Maldives, Myanmar, Thailand, Malaysia, Indonesia, Singapore, Australia, Kenya, Mauritius, Seychelles and Tanzania): *Milan 2014* was a pivotal element in India's large-scale, declared and performed commitment to presenting itself as a regional naval power. It also fulfilled demonstrative and symbolic functions, given that "the symbolism of Milan was reinforced a few week later with the announcement by India's National Security Advisor, Shiv Shankar Menon, that the Indian Ocean island states of Seychelles and Mauritius would be joining India's existing naval arrangement with Sri Lanka and the Maldives" (Brewster 2015: 54–55). *Milan 2014* and the naval defense group called *IO-5* may be defined as "performances of power" and they are the clearest examples of India's engagement on the regional chessboard as a big player. Mohan traces this strategy back to the 1990s when "phrases from 'Aden to Malacca' or 'Suez to the South China Sea' were re-injected into the national security discourse" (Mohan 2010: 7): This strategy was accompanied by infrastructural and military investment (e.g. equipment, training, and financial support) in ports located in archipelagic states and territories across the IOR such as the Maldives, the Seychelles, and Mauritius. The defense of oceanic space became a key national security narrative. According to Shukla (2012), a leading outcome of this strategy, with implications for regional geopolitical equilibrium, has been the establishment of Indian naval supremacy at multiple sites along the major important shipping lines. This supremacy is well exemplified, and publicly performed, by the reinforcement of Indian military presence in the Lakshadweep and Andaman archipelagos.

The increase in military presence and military control over oceanic chokepoints (especially along the Nine-Degree Channel between Lakshadweep and the Maldives) may be viewed as one of the main factors in the so-called "Indian Ocean Dilemma" (Kaplan 2009, 2010; Brewster 2015, 2017), namely the rivalry between India and China, and the potential conflict between their regional security policies. In recent decades, China has justified its policies by claiming the need to ensure the passage of oil tankers along maritime routes crossing the IOR. In 2010, Paul published a detailed analysis of the role of energy security in Chinese foreign affairs, forecasting that by 2030 some 29% of national energy demand will be met by oil, almost half of which will be imported from Middle East. The sea routes of the IOR remain the cheapest and safest option for the transport of oil. As already noted, these routes cross regional space, passing through chokepoints including Hormuz, the Nine-Degree Channel, and the Malaccan Strait, which have come to play a key role in satisfying the energy requirements of mega-industrial producer China. In the context of a more general strategy of self-affirmation as a global power, over the past decade, the Chinese government has been working on a Maritime Silk Route Initiative (MSRI), that is to say, the establishment of a safe passage across the ocean, formally called for to guarantee the safe transport of oil and goods. The MSRI relies on the economic influence that China can bring to bear over a set of strategic nodes known as the "String of Pearls". With the declared goal of counterbalancing Indian naval supremacy, China, acting through national companies, has led, announced, or been involved in the construction or restoration of hard-infrastructures, harbors, bridges and free trade zones, located at these nodes—a set of assets including ports, districts and islands situated at key points across the ocean space (Gwadar in Pakistan, Colombo in Sri Lanka, Kaafu in the Maldives). Palit emphasized the political (and geopolitical) function of MSRI, claiming that "these investments would give China control over the new infrastructure assets enabling their potential utilization for military and strategic purposes" (Palit 2017: 293). This interpretation reinforces the theory of growing Sino-Indian tension on the head of both countries' involvement in regional affairs. This tension has taken material form through the building of hard-infrastructures and military bases in archipelagic and coastal areas. Furthermore, these hard-infrastructural assets may be read as a performance of China's potential economic influence over the countries bordering on the Indian Ocean, and over the island and archipelagic states (e.g., the Maldives, Sri Lanka and the Seychelles).

Mohan (2010) has pointed out that India played a key role in the post-colonial period by challenging the construction of "foreign bases" across the IOR, lining up against the rise of a single big player following the end of British hegemony. However, "the Unites States has been the predominant power in the Indian Ocean since the early 1970s. The United States found itself compelled to gradually step up its military presence in the Indian Ocean as Britain withdrew" (Brewster 2017: 273): Subsequently to this phase, the United States' position on the regional chessboard has been described as in "elegant decline" (Kaplan 2009, 2010), or as undergoing progressive rearrangement in response to the rise of China and India as regional powers. However, as briefly mentioned earlier, the military base located on Diego Garcia continues to represent a key asset for the USA, in the context of its geopolitical involvement in the region. Furthermore, Doyle has reminded us that "the notion of the declining power of the US needs to be further developed" (Doyle 2016: 64) in relation to changing global strategies and the transition from a "traditional geostrategic game [to] a more flexible" (*ibidem*) one.

In sum, one of the key points we wish to make here is that recent decades have seen the (evolving) consolidation of a triangular condominium led by these three big players. These actors have reinforced their own positions within the regional scenario by reproducing three geopolitical narratives: maritime security and the protection of coastal regions in the case of India; energy security in the case China; and the legacy of the role of global security guarantor taken over by the USA following British disengagement. We need to be aware that these narratives interact with the geography of the region. Indeed: "major strategic actors have long sought to use geographical constraints to maintain the Indian Ocean and its littoral as a relatively closed strategic space". (Brewster 2017: 270): The control of access points, as well as of the main nodes and lines of communication, are geographical features that still shape the balance of geopolitical power in the IOR. The security narrative may only be read if we understand the geography of the ocean's spaces and the strategic position of certain archipelagic regions.

4.4 Key Archipelagos: Chagos and the Maldives

In 2015, Baruah labelled the small islands as the "holder of the key" to the Indian Ocean. Baruah's statement may sound hyperbolic; however, as discussed in the previous paragraph, a small number of archipelagos (forming states or other kinds of jurisdiction), namely Chagos, the Maldives, the Seychelles, the Lakshadweep, and the Andaman, occupy strategic positions in the context of regional maritime security. The literature defines these actors as "weaker" or "smaller" players, thus emphasizing their contrast with the "big" powers leading the triangular condominium and, at the same time, implying that these states and jurisdictions are of marginal or minor geographic relevance. This contrast makes the study of their geopolitical function even more relevant for our purposes.

The historical trajectory of Chagos shows us that, despite the transition to a post-colonial system of governance (BIOT), some key sites in the Indian Ocean continue to function as strategic platforms, on which to claim, or defend, a position of supremacy over oceanic spaces. This has given rise to a debate among experts and scholars on the meaning to be attributed to post-colonial institutions, especially in the regions and territories formerly under the control of the British Empire.

The Maldives' position on the fringe between China and India is even more exemplary. On the one hand, the archipelago comes under the "umbrella" of Indian military supremacy: for example, through its involvement in the *IOR5* group. On the other hand, some areas of the country have been converted into strategic assets via the investment of Chinese companies, as in the case of the Sinamalé bridge. The Maldives can play a so-called "Chinese card", and, at the same time, observe their naval military alliance with India. In other words, they can implement policies that support and ease the economic penetration of Chinese companies into their free-trade zones or stipulate agreements for the infrastructural strengthening of the country, for example by building marinas or by renewing or enlarging ports and maritime commercial facilities, without running the risk of losing the protection of Indian navy. In these insular spaces, the rivalry between the two powers plays out in terms of differential geo-strategies that generate the opportunity, for the smaller and weaker states, to attract foreign investment.

Chagos is a peculiar case that is not fully representative of the transition from colonial to post-colonial regimes. The role of the BIOT as a post-colonial institution deserves more in-depth analysis and, even more crucially, the function that the archipelago plays as a military outpost guaranteeing US dominance in the region is in keeping with a Mahanian perspective on the balance of power in the Indian Ocean. The case of the Maldives, on the other hand, shifts the focus onto the advantages and disadvantages of oscillating within the "fringe" that is an inherent feature of intensifying political and economic competition between the two giants on the regional chessboard.

Notably, against the backdrop just outlined, the development policies of island states, and more generally of the group of states that border onto the Indian Ocean, have become a key focus within the geopolitics of the region. This is certainly not a matter of "role reversal". Indeed, as we have documented, the triangular condominium still "holds the keys" to the ocean; nonetheless, scrutiny of small island development is certainly one of the factors to be included in a broader geographical and political framework for the IOR.

4.5 Future Scenarios: Geopolitics of Change

This geographical reading may be enriched by the introduction of an additional set of themes. As earlier stated, regionalization is first and foremost a geo-writing process that includes, for example, the choice of a scale to frame a region. This choice is far from being a neutral operation, and it strongly influences the interapretation of power relations among the actors. In this paper, we have cited the scale adopted by Rumley et al. (2012).[5] Chacko (2016) highlighted the emergence of a narrative that frames the Indo-Pacific

[5]As reminded, they proposed to adopt either a formal criterion to include a group of states bordering on the Indian Ocean, or a functional one, comprising all "states that have an interest in the maximisation of the Ocean's security" (Rumley and Doyle 2015: 10).

as the central stage of the twenty-first century, focusing on the bilateral relationships between China and the USA. The prevalence of this macro-region in international relations and geopolitical studies would reinforce China's global role, indirectly marginalizing the players that have invested in naval supremacy across the IOR: hence India.

Security does not only concern energy supply or the transit of goods. On the regional scale, environmental security is a priority for the big players and other international actors (such as international organizations and global NGOs). Furthermore, the blue economy (especially in relation to food security and sustainable fisheries policies), and the social and environmental threats associated with climate change rank high on the political agendas of Small Island States. Rumley (2010) referenced the discourse on climate change in his analysis of geopolitics in the IOR, and Doyle has defined climate change as a "conflict multiplier" (2016), observing that it encompasses environmental, political, and social challenges. The strengthening of the Indian navy's commitment may also be read as a strategy to ensure security for coastal regions and archipelagic areas (Gadihoke 2010). Furthermore, we must remember that the islands that "hold the keys of the IOR" are, geographically speaking, coralline atolls, whose physical characteristics include a high rate of erosion and which are seriously threatened by large-scale stresses to coral reefs (such as human impact caused by tourism practices, the acidification of ocean, or large-scale anomalous bleaching phenomena).

Special Economic Zones (SEZ) and free-trade zones may be viewed as a "weapon in the hands" of the archipelagic states. These economic institutions may reinforce the attractiveness of insular spaces for foreign investors. Such a process bears highly significant implications for the governance of marine areas and environmental protection policies. Moreover, it becomes a geopolitical factor for "coral states" (such as the Maldives) who are likely to be faced with future losses of terrestrial territory, and, consequently, may view investment in hard infrastructures across maritime spaces as pivotal to their economic development.

These considerations prompt a key question: do power imbalances among small islands and the players of the triangular con-dominium constitute a new form of imperialism? Recently some authors have debated this hypothesis (Paul 2010; Frost and Murray 2015; Blanchard 2017; Blanchard and Flint 2017). Here, we wish to point out that the archipelagic regions continue to represent strategic platforms offering economic and military control over the IOR, and that few noticeable changes have occurred since the shift to the post-colonial regime. This centrality, however, cannot be understood if we focus exclusively on the geopolitical chessboard; it also implies, as we have mentioned, awareness of environmental policies and interactions among a complex network of actors at the regional scale.

References

Agnew J (1998) Geopolitics re-visioning world politics. Routledge, London

Agnew J (2000) From the political economy of regions to regional political economy. Prog Hum Geogr 24 (1):101–110

Alpers EA (2014) The Indian ocean in world history. Oxford University Press, Oxford

Australia India Institute (2013) Task force on Indian Ocean security (2013): The Indian Ocean region: security, stability and sustainability in the 21st century. Melbourne: Australia India Institute

Baruah DM (2015) The small islands holding the key to the Indian Ocean. The Diplomat, 24 Feb. 2015. Retrieved from https://thediplomat.com/2015/02/the-small-islands-holding-the-key-to-the-indian-ocean/

Blanchard JMF (2017) Probing China's twenty-first-century maritime silk road initiative (MSRI): An examination of MSRI narratives. Geopolitics 22 (2):246–268

Blanchard JMF, Flint C (2017) The Geopolitics of China's maritime silk road initiative. Geopolitics 22 (2):223–245

Bouchard C, Crumplin W (2010) Neglected no longer: the Indian Ocean at the forefront of the world geopolitics and global geostrategy. J Indian Ocean Reg 6(1):26–51

Brewster D (2015) An Indian Ocean dilemma: Sino-Indian rivalry and China's strategic vulnerability in the Indian Ocean. J Indian Ocean Region 11(1):48–59

Brewster B (2017) Silk roads and strings of pearls: the strategic geography of China's new pathways in the Indian Ocean. Geopolitics 22(2):269–291

Chacko P (2016) India and the Indo-Pacific from Singh to Modi: geopolitical and geoeconomic entanglements. In: Chacko P (ed) New regional geopolitics in the Indo-Pacific: drivers, dynamics and consequences. Routledge, London, pp 43–59

Doyle T (2016) Climate change ad a comprehensive security in the continuum: geostrategy and geoeconomies in the time and place of the Indo-Pacific. In: Chacko P (ed) New regional geopolitics in the Indo-Pacific: drivers, dynamics and consequences. Routledge, London, pp 60–73

Frost T, Murray CRG (2015) The Chagos Islands cases. The empire strikes back. In: Northern Ireland Legal Quarterly 2015. Retrieved from https://ssrn.com/abstract=2649387

Gadihoke N (2010) Climate change: implications for the Indian Navy: maritime affairs. J Nat Maritime Found India 6(1):116–131

Jeffery L (2007) How a plantation became paradise: changing representations of the homeland among displaced Chagos islanders. J Roy Anthropol Inst 13:951–968

Jeffery L (2010) Forced displacement, onward migration and reformulations of 'Home' by Chagossians. J Ethnic Migr Stud 36(7):1099–1117

Jeffery L (2013) 'We are the true guardians of the environment': human-environment relations and debates about the future of the Chagos Archipelago. J Roy Anthropol Inst 19:300–318

International Court of Justice https://www.icj-cij.org/en

Kaplan RD (2009) Center stage for the 21st century. Rivalry in the Indian Ocean. Real Clear Politics, 16 Mar. 2009, Retrieved from https://www.realclearpolitics.com/articles/2009/03/rivalryintheindianocean.html

Kaplan RD (2010) Monsoon: the Indian Ocean and the future of American Power. Random House, New York

Mohan CR (2010) India and the changing geopolitics of the Indian Ocean: maritime affairs. J Nat Maritime Found India 6(2):1–12

Tuathail GÓ (1996) Critical Geopolitics. University of Minnesota Press, Minneapolis

Tuathail GÓ (2005) Geopolitics. In: Atkinson D et al (eds) Cultural Geography. A critical dictionary of key concepts, Tauris, London, pp 65–71

Palit A (2017) India's economic and strategic perceptions of China's maritime silk road initiative. Geopolitics 22 (2):292–309

Paul JM (2010) The role of energy security in China's foreign policy: a maritime perspective: maritime affairs. J Nat Maritime Found India 6(2):49–71

Pearson M (2003) The Indian Ocean. Routledge, London

Rumley D (2010) Ideology, carbon emissions and climate change discourses in the Indian Ocean Region. J Indian Ocean Region 6(2):147–154

Rumley D, Doyle T, Chaturvedi S (2012) 'Securing' the Indian Ocean? Competing regional security constructions. J Indian Ocean Region 8(1):1–20

Shukla A (2012) Indian Ocean: the new Great Game, Broadsword, 12 Sept. 2012. Retrieved from http://ajaishukla.blogspot.com/2012/09/indian-ocean-new-great-game.html

The Edinburgh Geographical Institute, John Bartholomew & Co (1922), Indian Ocean on Mercator's projection, The Times atlas, London

Do Not Cross. The 'North/South' Divide: A Means of Domination?

5

Pascal Clerc

Abstract

This paper would like to interrogate the concept of 'North', the concept of 'South' and the binary pattern which associates them: 'North/South'. These two concepts, developed during the seventies and eighties, seem to be more neutral than other terms such as 'underdeveloped countries' or 'rich countries'. But actually, it appears as a way to naturalize a division of the world in two categories of countries. This division, very stable for almost 40 years, is not a scientific way to think about the world. It is not a statistic division but more a heritage, and a vision of the world as immutable even against the facts.

Résumé

Cet article interroge le concept de "Nord", celui de "Sud" et le modèle binaire qui les associe: "Nord/Sud". Ces deux concepts, développés durant les années 1970 et 1980, paraissent plus neutres que les autres termes habituellement utilisés comme "pays sous-développés" ou "pays riches". Mais en fait, ces concepts permettent de naturaliser la division du monde en deux catégories. Cette division, très stable depuis 40 ans, n'est pas une manière scientifique de penser le monde. Ce n'est pas une division statistique mais un héritage qui renvoie à une vision binaire, immuable en dépit des faits.

Resumen

Este articulo interroga el concepto de "Norte", lo de "Sur" y el modelo binario que los asocia: "Norte/Sur". Estos dos conceptos, desarrollados durante les años 1970 7 1980, parecen más neutrales que los otros termos utilizados usualmente, como "países su-desarrollados" o "países ricos". Sin embargo, efectivamente, estos conceptos permiten de naturalizar la división del mundo en dos categorías. Esta división, muy estable desde 40 años, no es una manera científica para pensar al mundo. No es una división estadística sino una heredad que reenvía a una visión binario, inmutable a pesar de los hechos.

Keywords

'North/South' · Development · Representations · Domination · Spatial pattern · Neo-colonialism

Mexico City: 'the largest city in the world'. It was my first long trip at the beginning of the eighties. Innocently, I had dreamed of a radical otherness, of a coloured and exotic place, of apparent poverty everywhere, of strange plants, of new odours… I had dreamed of the 'South'. Obviously, it was a disappointment. At first, Mexico looks like other towns I was familiar with. I needed some time to exceed this disappointment and to discover more

P. Clerc (✉)
Laboratoire EMA, CY Cergy Paris Université, Cergy-Pontoise, France
e-mail: pascal.clerc@u-cergy.fr

B. Schelhaas et al. (eds.), *Decolonising and Internationalising Geography*, Historical Geography and Geosciences, https://doi.org/10.1007/978-3-030-49516-9_5

precisely differences and similarities with other places. Then, I continued my learning and my experiences of the world: the immaculate downtown of Saigon, the 'bobo' districts of Rio de Janeiro, the decrepit houses of North Queen Street at Belfast, the misery of Los Angeles downtown … It was less simple and less well ordered than I had originally thought. There were not two boxes for two worlds. The 'South' I had dreamed exploded. It was actually a complicated, heterogeneous, interlinked and contradictory set of spaces. Obviously, my idea of the 'North' was upset too.

It took a long time, readings and travels, for me to overtake this binary and simplistic vision of the world. It took a long time too to try to understand the uses of such a simple pattern, to understand its stability and its effects on the representations of the world.

This paper interrogates the concept of 'North', the concept of 'South' and the binary pattern which associates them: 'North/South'. Firstly to present the forms and the supports of them, than to explain their origins and finally to propose hypotheses about the stability of this pattern. My main hypothesis is that, behind these concepts that seem neutral, this binary pattern is a way to naturalize and determine a vision of the world.

I would like to add that this reflexion is done from a French point of view. Yet, different discussions and readings convinced me that the 'North/South' pattern was not only a French vision of the world even if it takes specific forms in a country of the 'North' which is strongly affected by its colonial history in the 'South'. But a part of the international literature about this question shows that the 'South' and more specifically 'The global South' could be a concept of repossession of their history, their identity and their space for countries of the former non-aligned Third World (Sparke 2007; Mignolo 2011).

5.1 A Structuring Spatial Pattern

The first part of this text is a kind of inquiry whose goal is to measure the importance of the 'North/South' model as represented in French public debates. For that, I analyze three kinds of public communication: media, political discourses and school. Each one is producing and revealing main representations. I will limit myself to the main conclusions of this inquiry.

Regarding the media, the 'North/South' model seems very important but this importance is varying depending on events. A survey of the French newspaper *Le Monde* shows clearly these changes between the mid-1970s and 2017.[1] Three periods appear as the subject of many articles: the first is the beginning of 'North/South dialogue' from 1975; the second starts with the publication of the Brandt Commission report in 1980; the third is characterized by the more recent diplomatic meetings on climate change (Fig. 5.1). We can cross the two last subjects: on the one hand development inequalities, on the other negotiations about climate change. These issues link 'Southern' countries with different situations: assistance, confrontation and dialogue. These situations have been the two forms of 'North/South' relationship from more than 40 years. They are the way to strengthen and to regard as evidence a binary division of the world. The same is also done by another French newspaper, *Libération*, the organ of the Socialist Party, which publishes almost every day a brief that presents, and sometimes opposes, information about a country of the 'North' and another one of the 'South'.

I took the opportunity of the French presidential elections of May 2017 to analyze the representations of the model 'North/South' in the French political world. I did this by exploring the websites of political parties and the official material[2] diffused by the candidates at the occasion of the first round of the election. Most of the

[1]The number of occurrences is huge. I decided to limit my corpus to papers containing the words 'North' and/or 'South' in the title. I analyzed 390 papers.

[2]In France, each elector receives an electoral circular (also called 'profession de foi') for each candidate. It is a document that presents the ideas, values and projects of the candidates.

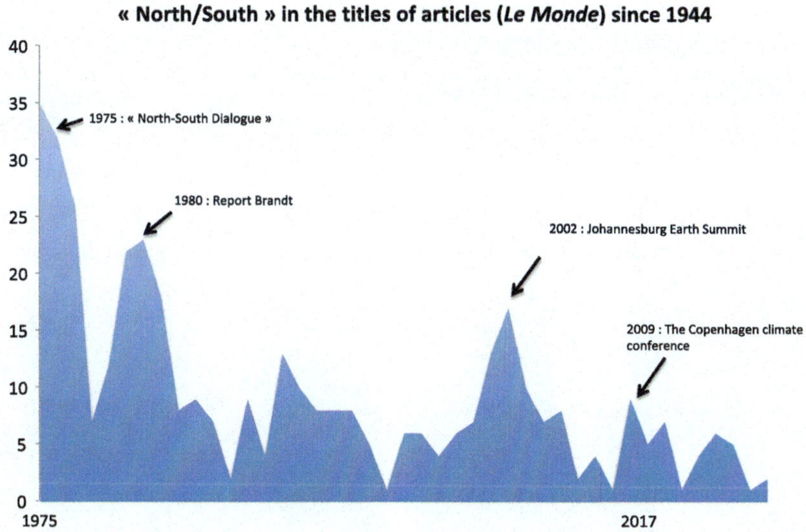

Fig. 5.1 This graph shows the evolution of the use of the term "North/South" in the titles of articles in the French newspaper Le Monde. The expression was never used before 1975. It becomes frequent in a period of crisis when it comes to reporting on the North-South dialogue and in particular the report of the Brandt commission. In the 2000s, the expression "North/South" was again used frequently to mention environmental issues

ideas exposed there involve the construction of a difference between France and the rest of the world. But the 'North/South' division is likewise evoked. In this case, the 'South' appears as a threat (very clearly in the Far-Right propaganda), or as part of the world particularly deprived and one which should be defended. A reflection on these political debates cannot be done without mentioning the speech that the then President of the Republic, Nicolas Sarkozy, made in Dakar in July 2007. Even if Sarkozy's remarks did not concern the totality of the 'South' but only Africa, we can identify there a representation of the 'African culture' based on very old stereotypes. On the one hand, Sarkozy denied the history of this region; on the other, he reduced this history to very coarse colonialist schemes like those which are referred to childhood, cyclical time or barbarity (Chrétien 2008). More generally, this text shows a representation of the 'South' as a form of irreducible otherness.

Finally, in the field of school teaching, the division 'North/South' appears as a structuring element of all the discourses developed in the texts or the iconography of the textbooks and in the official curricula that the teachers must follow.[3] Thus, texts, photographs or maps are frequently organized around binary presentations, which are considered as necessary to reduce the complexity of the world (Fig. 5.2). In this case, it is not even necessary to refer to the model 'North/South' because all this material embeds the model. However, with the last generations of textbooks (since the years 2010), one can notice an epistemological break with the most obvious markers of a separation between a 'North' and a 'South'. Thus, the famous 'North/South' limit presented in many world maps of textbooks since the 1980s, is more and more frequently discussed, even contested or erased in recent books. If this evolution shows some receptivity of the school to epistemological debates (which dispute the relevance of a 'North/South' limit), it does not erase the power of the simplifying construction that divides the world into two parts.

[3]It is worth noting that the French educational system is heavily centralized: primary and secondary level teachers have little freedom in choosing their teaching's topics and, for that, any view of the world imposed by Ministerial guidelines is susceptible of indoctrinating millions of young French students.

Source : Ivernel, M. (dir.) (2010) Histoire-Géographie, 5ème, Hatier

Fig. 5.2 These pages of a geography textbook for 12-year-olds present a discursive construction which is typical of French textbooks. It places two situations in binary opposition, here two learning situations. Without nuances, she confronts poverty and the lack of equipment in a "southern" country to a pleasant and well-equipped place of education in a "northern" country

This brief analysis of the media, political and academic communication show the primacy of a binary and essentialist representation of the world. Various activities in other national contexts than France convinced me that the division of the world between a 'North' and a 'South' was not limited to my country. Nevertheless, because of the specific history of France, we can think that this representation of the world may be more important than elsewhere. France was a colonial country where relations of domination over colonized people were based on representations of the 'Other' as inferior as a justification for domination. It is probably even more important to note that, in France, decolonization remained unfinished business, and it was never really accepted or conceptualized. The colonial thought persists with the existence of Overseas France or with current representations of migrants. In this context, the speech given in Dakar by an ex-President of the Republic was not a surprise. It expressed roughly an existing colonial difference, a gap between the two worlds. It is so much of an intricate and moving complex thought that the understanding of the contemporary world would require.

The 'North/South' model is a structuring spatial pattern. Besse (2004) and Besse et al. (2017) defines spatial patterns as structures that are both very global and very simple; they are an underlying basis of the explicit construction of knowledge, especially geographic knowledge. The 'North/South' model enters this category: a very crude superstructure often unthought, even invisible, but which determines the construction of knowledge about the world.

5.2 The Development of This Pattern

It is probably the British Oliver Franks (1905–1992), a diplomat and a civil servant for the World Bank, who first used the term 'North/South' in 1959 to qualify 'developed countries' and 'underdeveloped countries' in the following of Harry Truman discourse on January 1949. We could probably go back further with Alfred Sauvy and the 'Third World' or even at the colonial period with a hierarchy of the World ranging from 'civilized countries' to 'wild countries'. It could be interesting to explain what the imaginary of the 'South' is today. But Franks justified the choice of this term with a very plain remark: the so-called 'underdeveloped countries' are rather south of the earth, not necessarily in the southern hemisphere but south of the 'developed countries'.

Beyond the banality of this geographical observation, one wonders why did Franks chose these new terms. Perhaps, he wished to avoid stigmatizing term such as those who establish a hierarchy between countries. Actually, as I explain below, these seemingly neutral terms were stigmatizing too.

At the beginning, the term used by Franks had little success. It is the United Nations which will contribute during the 1960s to popularize the term 'North/South'. UN secretary between 1961 and 1971, U Thant, regularly highlighted tensions and inequalities between countries and used for that the term 'North/South'. He also contributed to give an institutional basis to the 'South' with the creation of the 'Group of 77'. The 1960s saw the birth of UNCTAD (United Nations Conference on Trade and Development), where a particular attention was given to these issues with the proclamation of the «United Nations development Decade». U Thant was a man from the 'South'; he was Burmese. He was the first UN secretary coming from this part of the world. It is sure that he uses the term 'North/South' to avoid stigmatization. But the road to hell is paved with good intentions.

It is in 1975 that the term «North/South» evolves. This new vocabulary appears during a succession of international meetings linked with the first oil crisis. On one side there are the 'wealthy countries' (also called the 'big' in the media, which is quite astonishing) which are sensitive to the cost of oil, on the other side producing countries that defend their interest. The French press often speaks of 'producing countries' and 'consuming countries', that is those who have resources and those who use them. It means that producing countries (actually the exporting countries of the 'South') are just playing a role of provider for other countries and do not have the needs of oil. Words betray representations. Gradually, meetings between countries like Venezuela, Algeria, Saudi Arabia or Iran (which are joined by Brazil, Zaire and India) on the one hand, and European Economic Community, Japan and United States on the other hand, will be called as 'North/South conferences' or 'North/South dialogue' by the French press.

But the decisive moment for the popularization of the term 'North/South' and its representation on maps is the publication in 1980 of the book *North–South, a program for survival*. It is the report of a commission chaired by the former German Chancellor Willy Brandt. While this report is linked with oil issues, we must go back to 1977 to understand its precise origins. In November 1977, under the auspices of the UN, Willy Brandt formed a 'North-South Study Group' (or 'North-South Commission') with eight 'rich' countries: Germany, United Kingdom, Sweden, Netherlands, United States, Canada, Japan and France; and eleven 'poor' countries, including Chile, Indonesia, Algeria, Kuwait, Colombia, Tanzania and India.

The aim of the members of the commission was to change the economic and social organization of the world by acting in particular on the relations between the two groups of countries. The commission calls these two groups 'North' and 'South' in the title of the book but also 'rich countries' and 'poor countries' or 'developed countries' and 'underdeveloped countries' inside the book. Brandt and his colleagues pretended to take some distance from a simplified image of a world divided into two camps but actually they contributed to this simplified image, this simplified and binary world.

It is especially the map on the front cover of some editions (the report is translated into about 20 languages) which gives a new representation, a spatialized representation of the 'North' and the 'South'. The map gives a feeling of truth and transforms this division of the world, this cartographical and semantic creation into evidence. For the first time, 'North' and 'South' are shown and precisely delimited.

5.3 The 'Naturalization' of This Pattern

Through the media and the political or academic discourse, the division of the world between 'North' and 'South' seems to have settled as a common representation. Above all, beyond political, social or economic issues, it now concerns different fields as music, sports or cinema that develop the popular diffusion of this representation of the world.

This pattern was also consolidated through it 'naturalization'. I understand naturalization as the transformation of an intellectual construction into natural reality. The naturalization of the 'North/South' pattern places this representation before any reflection, analysis or discussion. It is impossible to deconstruct it. It is outside of the epistemological field, outside of a history. It becomes a fact, an evidence. For this reason, it takes place in a field of geography described as 'realistic' that would consider world as a reality and not as an intellectual construction (Orain 2009).

In which ways is this naturalization working? Like many iconographic representations, maps can turn constructions into facts. By the representation of forms, they give an evidence to representations; then the representation of reality becomes reality. The strongest act which spatialized the division between 'North' and 'South' was the publication of the report led by Willy Brandt. On the map of the cover of the French edition (Fig. 5.3), a thick line divides two groups of countries represented by two very recognizable colours. Its meaning is unambiguous: there are two worlds in the world. This map, on the cover of the book, indicates a point of departure and the framework in which the reflection will take place. This map is the basis of many other maps which will invade the media and textbooks with, most often, a legend that clarifies things by indicating what separates this limit. It is very rare to have authors of textbooks, cartographers or journalists who indicate that this kind of map is intellectual constructions and who indicates the context of their creation.

Let us come to words now. First of all, one might say that nothing changes with the vocabulary used. This is partly true. Before the mobilization of the 'North/South' definition, we spoke of 'rich' and 'poor', 'developed' or 'underdeveloped' to designate almost the same sets of countries. So what do the words change if this binary division is not a novelty? Before 'North' and 'South', the terms referred priority to an economic and social situation. It established a form of hierarchy between places in the world and designated some of them as a reference: on the one hand, a group of 'developed', 'rich', 'industrialized' countries, on the other the rest of the world. The development western worldview defines a standard, a goal to be achieved and a possible point of arrival. Terms used to refer to the rest of the world say the gap to this standard: these countries are 'non-industrialized' (or 'under-industrialized'), 'underdeveloped' (or 'developing' to not to despair excessively).

It is probably to overcome this pejorative terminology that the actors mentioned above have used 'North' and 'South', two cardinal points that indicate a direction or a relative position and seem to be part of a more neutral vision of different territories. These words have replaced history by space: the history of a

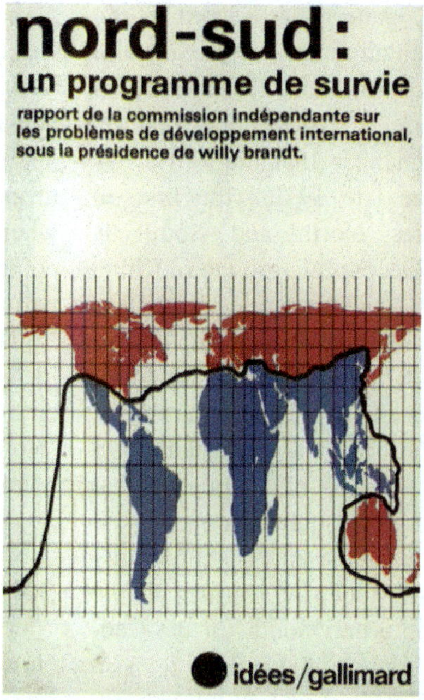

Fig. 5.3 The cover of the French edition of the Brandt report presents a world map which is shared by several editions (notably in English and German). This map uses a projection known as of "Gall-Peters". It is a little used projection which, beyond the technical aspects, conveys a political message and highlights the importance of the intertropical zone. But this message is scrambled by the line that clearly separates "North" and "South"

Source :
https://www.librairiedialogues.fr/livre/118344-nord-sud-un-programme-de-survie-gallimard

possible access to development, wealth and industrialization by a location. With this change, words contributed more than before to naturalize this binary representation of the world: if the question of development can be thought as a possibility of change, the location in the world does not change. It is possible to have control of its development, not of its location. The novelty of this identification with cardinal points is that it is impossible to change. It is more discriminating, more exclusive than the older denominations. Indeed, a 'poor country' can hope to become rich, an 'underdeveloped country' can hope to become developed, a 'developing country' seems even on the way. There is a potential movement, perspectives in these designations, even when they are thought by the western worldview. The 'North/South' pattern shows a world order; it says where are the ones and the others are. It establishes a spatial hierarchy with the 'North' at the top and the 'South' at the bottom. However, there are counterexamples as Australia and New Zealand that a cartographic sleight of hand has ranged from the Brandt Report to the 'North', as North Korea, which is in the 'South', and South Korea, which is in the 'North'. Nothing is ever perfect.

But generally, a country located in the South remains there. Moreover, more than 35 years after the Brandt report, and despite major geopolitical and economic upheavals, the map remains surprisingly stable. Probably because to question it, is questioning the world order. Let me add another element: to mobilize these two cardinal landmarks to define countries can also mean a form of submission to climatic determinism: in the 'South', the climate is hot (even if it is not always true) and implicitly, it can allow to update all the old strong clichés relating to the influence of climate on the development of societies which are used from Montesquieu to the colonialists.

Based on this division of the world are constructed speeches and categories from which statistics are produced. And these statistics help to give an identity to these entities. The name creates the category and creates an appearance of reality. 'Naming is creating' as the French geographer Raison (2007) wrote.

The 'North/South' pattern has settled in the speeches and representations for 40 years and since that time, it has not moved. This stability is the main indicator of its preconceptual nature. Indeed, despite the changes that the world has experienced since the late 1970s, the line on maps, which separates 'North' and 'South' is unchanged. The quick economic growth of China and many Southeast Asian countries, the stagnation of Europe, the fragmentation of the USSR, the 'normalization' of Latin America, the entering into a form modernity of the oil monarchies, the questioning of the triad and then of the US hyperpower are some of the developments that have upset the major global balances.

But a careful look at the maps shows only three minor changes: Israel's attachment to the 'North', integration into the 'South' of the Caucasian states and Central Asia (but it is a geopolitical effect of the end of USSR), integration into the 'North' of Singapore. That's all.

Finally, how can we define the countries of the 'North'? And of the 'South'? If these two entities have a meaning, they must be measurable. It is possible to make a test with two very common indicators: the Human Development Index (HDI) and the Gross Domestic Product per capita. Logically, if we classify all the countries of the world according to one of these criteria, the last country of the 'North' should precede the first of the 'South'. But even leaving aside a few small states whose situations are very specific, the limit, which appears on maps and in the vocabulary, seems to be an illusion. Some examples: Chile, a country of the 'South', with an HDI of 0.822 (40th worldwide) is far ahead of Serbia (0.745 and 77th worldwide) or Ukraine (0.734 and 83rd worldwide) traditionally classified in the 'North'; Malaysia, another country of the 'South', has a GDP per capita of $ 24,714 in 2014 (41st worldwide), just ahead of a European state (so-called 'North'), Hungary, and far ahead of another, Macedonia, whose GDP/inhabitant is half and 79th worldwide. I could multiply the examples: with these two indicators or others of the same nature, it is simply impossible to draw this limit. In this logic of deconstruction, let us also mention the low interest that it has, in a globalized and complex world, to divide the space in such a rudimentary way, relying moreover on a surface rather than reticular apprehension of the space. One could also note the demonstrative weakness of a division that organizes the world at the state level while everywhere, in the richest countries as in the poorest countries, regions, cities, neighbourhoods, prosperous and other disinherited people appear.

Therefore, the 'South' appears more as an idea and a heritage than as a statistical category, an idea constructed by political actors of the 'North' and largely integrated, especially in the countries of the 'South'.

5.4 A Resilient Pattern

How can we explain that this model resists despite its limits, its bad scientific relevance, its inability to give meaning to the contemporary world and despite the criticisms of which it is the object? The purpose of this last part is to analyze the causes of this resistance.

We can first identify a kind of reflex, the desire to insert a vision of the world in a reference which seems stable. The 'North/South' pattern organizes the world and it does it in a simple and reassuring way. The world divided between 'North' and 'South' is a world where nothing changes: the 'South' is still in the south and this distinction refers to one of the oldest spatial patterns: a zonally one of a climatic inspiration (Besse 2003). This pattern constructs for the 'South', an otherness which is so far to Western civilizations.

Within the complexity of a globalized World, this binary division reassures and puts an appearance of order in the world by distinguishing 'Us' and 'Them' (Dufour 2007; Hancock 2007).

The line between 'North' and 'South' also divides two entities that would sometimes be ontologically different, sometimes put in comparative tension, with a 'North' as a reference for a 'South' whose target would be to close this gap. Jullien (2012) shows how this increasing distance plays both comparison and difference, and how these two ways of apprehension of otherness cannot be separated from a relationship of domination. In this context, it seems impossible to develop an understanding empathy and to study other kinds of coherence. Jullien prefers the figure of the 'gap' because it allows to discover the specificity of the otherness and its own coherence. With the 'North/South' pattern, the 'South' is reduced to its radical otherness, its strangeness and its developmental delays.

The existence of the 'South' therefore allows the countries of the 'North' to define themselves by opposition. In 'North' and 'South', there is an idea of top and bottom (even if it has not always been so on maps), there is the head and the rest of the body, there is thought and corporality, rationalization and emotion. It is once more the reference of the infamous Dakar speech that shows a 'Northern' leader mobilizing most of these couples of opposites. On the same topic, I would like to mention a violent and ironic text, but in my opinion very lucid, of the philosopher Paul B. Preciado. He shows how, why and on what basis, the 'North' created the 'South':

«le Sud est primitif et passé. Le Nord est progrès et futur. Le Sud est le résultat d'un système racial et sexuel de classification sociale, une épistémologie binaire qui oppose haut et bas, l'esprit et le corps, la tête et les pieds, la rationalisation et l'émotion, la théorie et la pratique. (…) Le Sud est potentiellement malade, faible, stupide, incapable, paresseux, pauvre. (…) Le Nord se présente toujours plus sain, plus fort, plus intelligent, plus propre, plus productif, plus riche.» (Preciado 2019: 290)[4]

Finally, from a French point of view, a last question remains to be evoked: the colonial one. Based on the traumatic events of the Algerian War and on a problematic migratory context, France is constantly replaying colonial conflicts. The prevalence of the 'North/South' pattern in the media, school culture or political discourse can be analyzed as a legacy of this unfinished decolonization of mentalities. The 'South' is mainly constituted by areas colonized formerly by the Western powers. We can go further: through the vocabulary used, the stability of the dividing line, the mental representations or the aid to development, the richest countries invent a kind of neo-colonialism.

5.5 Conclusion

'On the wall opposite, there is a map of the world. She is upside down, Antarctica up. He realizes that I'm staring at her. "You are from the North," he said, "those in the North can not believe that their beautiful planet is upside down, but for us the world is like that, the South up." My eyes are lost on the map. (…) The card reversed right now seems to me, it teaches me to be on the antipode. The flight that I believed towards the bottom turns upward. I'm on the tip of a rock and I'm waiting for the dive'. (De Luca 1999: 117)

These words of the writer Erri De Luca (1999) remind us that we can look at the World differently and suggest we should do it. It is a first step towards a reflexive approach, a way of no longer considering the World and its divisions as a fact but as a construction, a meta-geography which makes possible to 'analyse the spatial structures through which people order their knowledge of the world' as written by Lewis and Wigen (1997: ix).

Along with the 'invention' of the continental division (Grataloup 2009), the 'North/South' pattern has become self-evident, as a precise and faithful translation of the reality, despite its extreme simplification of a complex, cross-linked and multiscale World.

The pattern 'North/South' reassures by organizing the links with otherness (even for the inhabitants of the so-called 'South'). It is also performative and acts on reality. We organize our actions from our representations. The pattern 'North/South' participates in the action of societies in the world and, as a self-fulfilling

[4]'The South is primitive and backward. The North is progress and future. The South is the result of a racial and sexual system of social classification, a binary epistemology that opposes high and low, mind and body, head and feet, rationalization and emotion, theory and practice. (…) The South is potentially sick, weak, stupid, incapable, lazy, poor. (…) The North is always healthier, stronger, smarter, cleaner, more productive, richer'.

prophecy, it builds what it has stated. For example, it builds a 'South' of all dangers and therefore produces very real, very material borders, such as between the United States and Mexico or between Europe and the southern shore of the Mediterranean basin. With development aid, it reinforces a process of domination of one part of the world over another.

The words are not innocent.

References

Besse J-M (2003) Les grandeurs de la terre. Aspects du savoir géographique à la Renaissance, ENS Éditions, Lyon

Besse J-M (2004) Le lieu en histoire des sciences. Hypothèses pour une approche spatiale du savoir géographique au XVe siècle. Mélanges de l'École française de Rome. Italie et Méditerranée 116(2):401–422

Besse J-M, Clerc P, Robic M-C (2017) Qu'est-ce que le «spatial turn»? Revue d'histoire des sciences humaines 30:207–238

Brandt W (1980) Nord-Sud, un programme de survie. Gallimard, Paris

Chrétien J-P (ed) (2008) L'Afrique de Sarkozy. Un déni d'histoire, Karthala, Paris

De Luca E (1999) Trois chevaux. Gallimard, Paris

Dufour F (2007) Dire «le Sud»: quand l'autre catégorise le monde. Autrepart 41:27–39

Grataloup C (2009) L'invention des continents. Larousse, Paris

Hancock C (2007) «Délivrez-nous de l'exotisme»: quelques réflexions sur des impensés de la recherche géographique sur les Suds (et les Nords). Autrepart 41:69–81

Jullien F (2012) L'écart et l'entre. Leçon inaugurale de la Chaire sur l'altérité, Galilée, Paris

Lewis MW, Wigen K (1997) The myth of continents. University of California Press, Berkeley, A critique of metageography

Mignolo WD (2011) The global south and world dis/order. J Anthropol Res 67–2:165–188

Orain O (2009) De plain-pied dans le monde. Écriture et réalisme dans la géographie française au XXe siècle, Harmattan, Paris

Preciado PB (2019) Un appartement sur Uranus. Grasset, Paris

Raison JP (2007) Nommer, c'est créer un peu. De «Tiers-Monde» à «Tropicalisme»: les avatars d'un vocabulaire. Autrepart 41:57–68

Sparke M (2007) Everywhere but always somewhere: critical geographies of the Global South. The Global South 1:117–126

Drone Photography and the Re-aestheticisation of Nature

6

Verónica C. Hollman

Abstract

Drones offer the possibility to depict places from perspectives difficult to achieve. They make visible an order of forms, patterns and relationships that, from the surface is either difficult or impossible to have. Although originally developed as a military technology, and hence highly restricted to military practices, drones are currently more accessible for a wider range of users. Perhaps, drone photography and its recent popularisation are the most outstanding evidences of how the use of this technology has experienced a shift from the military to the civil arena. Both, the altitude and the proximate perspective that drones achieve, make the drone picturing a renewed challenge. Assuming that each technology of vision organises what we observe with our eyes, I will explore how drone photography is picturing nature. By focusing on the visual universe displayed under the category Nature on Dronestagram, a website devoted to drone images, I will identify some of the mechanisms upon which drone photography is reshaping the geographical imaginations of nature.

Résumé

Les drones, en règle générale, permettent d'avoir accès à un type de vue aérienne rendant perceptible une organisation des lignes, des formes et surtout des relations spatiales qui ne l'était pas, ou peu, au sol. L'utilisation de ce type de dispositif, qui avait été développé, au départ, comme une technologie militaire, n'est plus limité à ces pratiques. Sans doute, le recours de plus en plus courant aux drones pour réaliser des photographies et des vidéos, qu'elles soient récréatives, publicitaires ou scientifiques, reflète la diffusion croissante de ces engins dans le domaine civil. L'altitude, d'une part, et la proximité, de l'autre, proposent de nouveaux défis à la photographie. À partir de l'idée selon laquelle chaque technologie de prise de vue organise ce que l'on voit et comment on le voit, je me demande comment les drones photographient la nature. J'analyse, plus précisément, l'univers visuel qui nous est proposé, dans la catégorie: «Nature», par le site web «Dronestragram», créé pour diffuser les images réalisées à l'aide de drones. Je propose d'identifier certains des mécanismes utilisés par ce type de photographie et qui sont en train de modifier considérablement les imaginaires géographiques de la Nature.

V. C. Hollman (✉)
CONICET, Universidad de Buenos Aires, Instituto de Geografia "Romualdo Ardissone", Buenos Aires, Argentina
e-mail: vhollman@conicet.gov.ar

B. Schelhaas et al. (eds.), *Decolonising and Internationalising Geography*, Historical Geography and Geosciences, https://doi.org/10.1007/978-3-030-49516-9_6

Resumen

Los drones brindan la posibilidad de ver los lugares desde perspectivas que en algunos casos resultaría dificultoso lograr de manera directa. Este dispositivo hace visible un orden en las formas así como relaciones y patrones que desde la superficie sería dificil o incluso imposible tener. Si bien los drones fueron originalmente desarrollados como una tecnología militar, su uso actual ya no se restringe a las prácticas militares. El creciente uso de drones para realizar fotografías y videos con diversos fines (recreativos, comerciales, publicitarios, científicos, etcétera) es sin duda una de las evidencias más notables de su utilización en prácticas civiles. La altitud y la perspectiva de proximidad de los drones proponen nuevos desafíos en la captura fotográfica. Asumiendo que cada tecnología de visión organiza qué y cómo vemos, en este capítulo exploraré cómo se fotografía la naturaleza con drones. Para ello, analizaré el universo visual que se despliega en la categoría Naturaleza de Dronestragam, un sitio de Internet dedicado con exclusividad a la difusión de fotografías y videos realizados con drones. Como itinerario de análisis identificaré algunos de los mecanismos a partir de los cuales la fotografía con drones está reconfigurando los imaginarios geográficos de la naturaleza.

Keywords

Drone photography · Aerial view · Nature · Geographical imaginations · Dronestagram

6.1 Introduction

Unmanned flying vehicles (UAVs), widely known as drones, were originally developed as a military technology. However, the increasing production of commercial drones is making its use more common for a wider range of civil practices, and photography is the most popular use (Rothstein 2015). The massive production of drone pictures and videos is a visual universe per se that largely exceeds the military origin of this technology. In spite of this shift to the civil arena, academic literature has problematised many more drones in the realm of warfare (Chamayou 2015; Gregory 2010, 2011, Gregory 2015; Kindervater 2015).

If we accept that each technology of vision organises what we observe, we can think that the popularisation of the drone, as a device, and the massive production and circulation of drone photos are reshaping our geographical imaginations. Among technologies of flight, drones enable us to have an aerial perspective from closer distances. While planes fly between 35,000 and 42,000 feet, balloons between 1000 and 3000 feet, in general, commercial drones should fly below 400 feet.[1] From this proximate distance, the drone offers either vertical or oblique perspective to see and picture the Earth. Commercial drones provide us with amazing images of canyons, glaciers, forests, mountains, deserts, oceans and wildlife. In other words, a fascinating visual experience that can feed the imagination.

In this chapter, I will focus on the visual universe that a specialised website—Dronestagram—displays under the category *Nature*. I will identify the mechanisms upon which drone photography is reshaping the geographical imaginations of nature. Drawing on this analysis I will suggest that some features of this genre of photography make viewers forget that drones are, as any form of the device created either to look at or to enhance natural vision (Mirzoeff 1999: 1), visual technologies that intervene directly in what is visible or not.

6.2 A New Bird's Eye View in the Civil Arena

Most of the literature within geography, as well as critical literature in social science (Chamoyou 2015), has focused on drones as a technology of warfare and military surveillance (Gregory 2011; Kindervater 2015). Briefly, research traced the

[1]Regarding drone's flight altitude, it must be said that regulations vary from country to country.

early development of drones in the UK during the 1920s, the 1930s and the 1940s as pilotless aircraft and offensive weapons of war with the aim of both extending the point of attack beyond human pilot's visual abilities (either under extreme meteorological conditions or during the nights) and reducing the risks to personnel in flying areas previously categorised as dangerous. With the Cold War, there was a re-emergence of drone development in the United States, as a tool for intelligence, reconnaissance and surveillance. The flying device became a tool to obtain photographs in areas with risky or difficult access. In the Vietnam War, the drone was used to obtain visual evidence of targets for air strikes and post bombing damage, as well as to distribute propaganda material and electronic listening devices.

The two most successful drone research and development programmes pursued by the US government in the 1990s—the Predator and Global Hawk—sought to increase drone intelligence, reconnaissance, surveillance and transmission capabilities. These programmes were put to the test in the late 1990s, particularly in the US and NATO air strikes over Kosovo. Post 9/11 2001, drone research has been oriented to the coming together of practices of surveillance and information gathering with those of targeting and killing. UAVs offer a promise 'of seeing, mapping and knowing a complex landscape' (Kindervater 2015: 117) upon which warfare has been organised during the last twenty years: 'they make it easier to kill than to detain because that killing remains largely hidden from (Western) view' (Kindervater 2015: 49). Perhaps the most relevant issue that these researchers point out is how the drones are reshaping warfare. Derek Gregory argues that the eyes of the new hunter-killers are now constructing particular bodies as targets in transnational hunter-killer missions as in Pakistan, Afghanistan or Somalia (Gregory 2010).

In tune with this literature, contemporary artists have introduced in their artworks the drone view as a topic of discussion. Among others, James Bridle's *Drone shadows* project consists of the shadow's drawing of the United States military drones in a one-to-one representation in different cities of the world. This artistic project makes visible what Derek Gregory has analysed as 'the emergence of a political technology of long-distance and stand-off killing (2015: 259).[2] The artist explains

> The Drone Shadow is not just a picture of a drone, it is a diagram of a political system. Every time we draw one, we use it to cast light on the actors who would prefer that the reality of their intentions and actions remain hidden. (Bridle 2014).

Whilst military use is still the common, it is estimated that drone markets will triple in value due to civil uses between 2015 and 2025. As an object, the drone 'is a transportation paradigm shift, it is a flying machine, it is composed of computational electronics, and it is an automated robot' (Rothstein 2015: 1). In other words, the drone becomes a suitable device for multiple practices within the civil realm. In fact, drone manufacturers have recently oriented their efforts to show that this object can be extremely useful for a plethora of civilian activities such as mail delivery, weather forecasting, crop monitoring, film industry, news reporting, security and police operations in civilian aerospace, architecture projects, many areas of scientific research and even for producing art. Israel Aerospace Industries, for example, have highlighted the need to discover 'ways to make people perceive drone technology as a natural part of a future society, to create positive interest in drones and to create a multidisciplinary promotional campaign (Graham 2016: 87). Open source drone manufactures, in a very similar way, are emphasising how drone technology can become a useful device to empower social movements and their fights.[3]

Commercial drones can be broadly categorised as camera drones, racing drones and toy drone, varying from US 30 to US 3000. More

[2]The app created by the visual artist Josh Begley also aims at making visible a political system of killing. The app sends to your cell phone a push notification every time a drone strike is reported in the news.

[3]Social movements and unions in Argentina frequently use drone photography as a document of the social support they have during protests and mobilisations.

sophisticated drones, throughout these three categories, include cameras stabilisation, software that opens more possibilities (such as gesture control), forward-facing, obstacle-avoidance, Wi-Fi connection and computer hardware for mapping out environments in three-dimensional space that help to avoid obstacles. By and large, they are becoming smaller and lighter to carry, and they are also achieving more time and distance flight autonomy. In general, these UAVs are multi-rotor or quadrotor which are much cheaper and easier to fly than helicopters. Only recently, the miniaturisation of multi-rotor technology has acquired more stability and autonomy of flight, and this change has a direct impact on the quality of the images obtained. The quadrotor (two sets of two propellers spin in opposite directions) is a relatively recent technology, and in the last few years, a number of companies have been selling full systems for flight photography, aerial mapping, and hobbyist uses.

Despite the increasing production and use of commercial drones, the academic literature has problematised much more the use of this technology in the realm of warfare. Regarding the use of UAVs in geographical research, Birtchnell and Gibson (2015) analyse the drone as a resource for teaching research. They identify the challenges of introducing drone technology, both in graduate teaching and social research: on the one hand, by reflecting on its legal and ethical implications and on how data is constructed with this technology; on the other hand, by imagining the topics of research that could benefit from its use. Garrett and Anderson point out that drones, as other devices in the history of science, 'enable us to extend our perception into new places, they multiply our possible experiences, and they reshape our geographical imaginations' (2018: 346). Hence, these authors emphasise the need to avoid conceptualising drones just as a tool for research by engaging in both the piloting of drones and its critical reflection.

Drawing on Rothstein's definition, I will state that the drones are also flying cameras that do not only enable the user to see places or phenomena but also to create still images or videos from the air-space above called Nephosphere (Garrett and Anderson 2018). Cheaper to have and easier to operate, there is no doubt that the commercial drones are also widening the experience of producing vertical and horizontal images. In fact, photography is one of the main current use in the civil arena (Rothstein 2015). The massive production of drone pictures and videos is a visual universe per se that largely exceeds its origin as a military technology. Suffice to say that one of the drone providers registered the production of more than 500 million pictures during 2016. In addition, the digital materiality[4] of these pictures promotes a wider circulation of the aerial perspective.

6.3 'Let's Build a World Map of Our Earth with a Bird's Eye View'

There are many websites devoted to the diffusion of drone pictures and videos. In 2013, Eric Dupin and Guillaume Jarret created Dronestagram 'with the aim of providing online resource for drone photographers'. Dronestagram emulates a social network with more than 35,000 users, mainly from the US, UK, Germany, Canada and Spain, where people share their drone images, as well as tips for obtaining better images. The website also offers information about photography contests and even hosts an International Photography contest.[5] In collaboration with Dronestagram, Ayperi Karabuda Ecer has recently edited a book *Dronescapes* with 250 drone photographs.

There are many itineraries available to the visitor through the site: winning photographs, suggested places, videos (reclassified by type of drone, quantity of visitors, likes and comments), photographs (reclassified by categories such as urban photos, nature, sports, social events, countries, dronies as well as by type of drone

[4]Digital technologies enable the visualisation of images through different digital devices (computers, smart phones, online distribution platforms) and different material qualities (hard copy, different screens). Regarding the geographies of digital technologies see Rose (2016).

[5]From 2014 to 2017, National Geographic as well as the main producers of commercial drones were sponsors of the International Drone Photography Contest.

used, the quantity of visitors, and the number of likes and the comments obtained). It is also possible to look at both the photographs and the videos by taking the world map as the initial point of a virtual trip. The map embodies the will to 'build a world map of our Earth with a bird's eye view'. The logo of Dronestagram appears all over the map and the user can zoom in and out to different places in the world that have been reached by the drone cameras. One interesting paradox emerges from the Dronestagram mapping: Iraq, Pakistan and Afghanistan almost appear as territories out of the drones' view despite the fact that they continuously experience US drone attacks. In other words, Dronestagram creates a geographical imagination of the world that excludes the fact that the population of those countries is being affected on a daily basis by military drones.

By defining the top places for tourism, Dronestagram organises the visual information of the world as a trip with global pretension from the Western perspective. The idea of making a trip through the visualisation of pictures recreates the lantern slide shows that were so popular by the end of the nineteenth century in the UK (Ryan 2011). However, the computer as the medium of visualisation reshapes the act of seeing as an individual experience. The organisation of global contests, as well as the publication of the winning photographs in well-known magazines, such as *National Geographic* clearly demonstrate Dronestagram's aims to reach a wide and global audience, not restricted to the community of drone photographers.

6.4 On the Production of Drone Photography

Dronestagram is frequently presented as a website for non-professional photographers. However, it is a site in which highly trained photographers publish their work. Their trajectories show that most of them were professional photographers before getting into the world of drone photography, and some of them have even set up companies devoted to drone image production. They have several drones (between 5 and 15 drones) with different characteristics and this means that they can choose the drone according to the places of capture, the weather conditions or the type of image they would like to obtain. The degree of professionalisation becomes evident throughout three areas of knowledge:

(a) The use of the drone as a flight device including the mastery of complex tasks such as carrying out manoeuvres and estimating the flight time depending on the autonomy of each device. This knowledge is essential to avoid losing or affecting the operation of the drone. One of the award-winning photographers states:

> Gradually I was learning drone. The more I was learning its opportunities the more complicated my flights were. I began to fly higher, faster, perform complicated manoeuvres, fight against nature. Soon I was able to launch it even from invisible point, more than a kilometre away from me. […] When I got used to control drone better I started to shoot for myself. At that moment I began to discover things. […] **Drone became more than just a device for me, it became my friend and "co-writer" for my stories**. Wherever I'm going it's always with me. (Maksim Tarasov, In: http://www.dronestagr.am/blog/interview-maksim-tarasov/ My underlying)

(b) Photography as a technology involves the practical knowledge of the cameras (their potentialities and limitations), as well as the visual training to identify accurate objects and situations to picture from the Nephosphere. Photographers, for example, frequently use Google Earth (another technology of vision) to search for the areas to picture from the drone. They also recall their desire to re-create an image they remember either from a previous flight experience or from aerial photographs they have observed before, suggesting that the visual memory is crucial in the production of drone photography. It seems that training the vision is as important as the training to operate the drones.

> Get a lot of practice but start with understanding the basics of photography. Just because it's with a drone doesn't mean you don't have to understand how light works, composition and editing (Jerome Courtiel, en: http://www.dronestagr.am/blog/interview-jerome-courtial/).

(c) Software for photo editing and video. Professional photographers recognise the digital edition as part and parcel of drone photography production. They even point out the use of more than one image editing software. In spite of this practice, Dronestagram's users frequently point out visual details supporting their suspicion of the digital edition of pictures.

In short, nothing is improvised or random in the production of the most viewed and liked photographs in Dronestagram. Drone photography involves training in the handling of a system of devices, previous survey of the area, an accurate selection of the place and time to take the photograph, as well as the drone to be used, the angle and the most appropriate point of view for each shot. The visual training of the photographer also shapes what and how they decide to picture. However, the production of drone photography goes on with the digital edition of the image, with software such as *Ligthroom* or *Photoshop*. Drone photography is the result of careful procedures that photographers carry out before, during and after shooting the photo. In the following section, I will analyse how all the drone is picturing nature.

6.5 Painting Nature with Drones

I would like to start by focusing on the three winning photographs in the category *Nature* of the Dronestragram 2017 contest, sponsored by the National Geographic magazine.[6] Two aspects called my attention from these three pictures, not individually but precisely in their dialogue. First of all, nothing in these photographs refers to a pristine nature. On the contrary, the three pictures depict highly transformed or produced natures. One of them depicts a lavender field that occupies the totality of the visual composition.[7] Another, shows a route tracing a sinuous shape in the midst of a forest.[8] The third one depicts from above multiple fragments of ice in the midst of a blue matrix, perhaps one of the most acknowledged effects of climate change.[9] Secondly, the beauty of the visual composition captures our eyes and provokes an emotional engagement. Beauty has definitely defined both the production and the selection of each of these photographs. The slogan of the contest sets a return to the classic correspondence between beauty and nature. However, both the perspective chosen and the organisation of the visual composition makes the social intervention and modification of nature as a *sine qua non* condition of its beauty.

'If it wasn't on Dronestagram, I would have thought it was a painting! Great job', writes one of Dronestagram's users. This comment suggests that drone photography provokes pleasure analogous to the one experienced by looking at a landscape painting. Perhaps one of the most outstanding features of drone photography is how colour contributes to the creation of this visual pleasure. As Philip Ball puts it, 'colour speaks to our feelings; however, culture shapes its meaning' (2013: 41) and compares its effects with music '[...] it comes by a shortcut to our senses and feelings' (2013: 15). Besides, this aesthetic pleasure the colour gives the viewer clues to identify shapes and the existence of different patterns. The colour also contributes to reinforce the homogeneity of a pattern and at the same time, it highlights its uniqueness. Taking as an example a drone photography of a field, the colour indicates areas with a homogenous land

[6]Stephanie Hawkins (2010) has analysed the impact of this magazine on the nation's visual literacy with a creative blend of scientific and aesthetic modes.

[7]First prize category Nature 2017 Contest: Provence, summer trim. Photographer jcourtial.

[8]Second prize category Nature 2017 Contest: Infinite route to Transylvania. Photographer Calin Stan.

[9]Third prize, category Nature 2017 Contest: Ice formation. Photographer Florian.

use but it also defines different types of crops within farming use. The colour can also change according to the precise moment in the crops' development, and therefore, it draws a temporal pattern.

Drone photography is the domain of digital colour. This means that the reflected light by the objects is recorded into a matrix of data instead of being reflected on emulsion paper. Digital programmes analyse the different parameters of the colour from data obtained. Photographers can edit the image by choosing to modify parameters such as hue, saturation and brightness. Hence, it is reasonable to state that much of the intensity and the brightness that characterise the drone photography is due to its digital processing. In addition to this digital production of colour, computer monitors, as one of the main spaces to exhibit drone photography, also affect the brightness of the colour. Paradoxically, the intensity and the high definition of the colour make audiences feel and believe that they are looking into reality.[10] The light blue of the ocean, the light purple of a lavender field or the white and green of a snowy forest are perfectly and beautifully composed with the outcome of the hyper-realism effect. To give another example, the first prize winner photograph of the 2015 contest, shows a couple of snorkelling in the midst of the light blue of an ocean.[11] The human couple and the sharks swimming nearby have that light blue ocean as a shared space. The colour gives transparency to the image and reinforces the realism of the scene: the viewer has the feeling of seeing that aquatic landscape without any kind of mediations.

Besides the colour, one of the winning photographers highlights that finding interesting shapes is one of the central tasks of drone picturing. Cosgrove and Fox (2010) suggest that finding and identifying shapes are activities directly associated with aesthetic pleasure. The perception of shapes is a structuring element of thought as it offers a degree of generality that allows it to be applied to more than one object and can be recognised with relative ease. Drawing on Cognitive Psychology, it is possible to understand that when we see an object we can only perceive it if we are able to refer it to an archetypal form (Arnheim 1997). This is why shapes are so crucial in the construction of the visual composition of drone photography. Interesting shapes means either recognisable or with a unique characteristic.

The three winning photographs of the 2017 contest, organise the visual composition around shapes, more or less regular but very defined. Even if nature's shapes are in general irregular,[12] the drone photography is particularly accurate to show the fitting of regular and irregular shapes as part of the same entity. In one of the winning photographs of the 2017 edition, for example, a sinuous route fits the regular pattern of the forest in such a way that it resembles a meandering river running through a wooded valley. The colours let the viewer know that the sinuous object is not a river but a route. However, the route's shape is so well defined and it fits so perfectly in the visual composition that leads the viewer to look at the route as the feature that makes that naturally beautiful.

As a result of taking a distance from the objects, drone photography changes the size, one of the attributes of the shape. In addition, it has two other effects on perception. On the one hand, the change in the angle of observation, more evident in the case of vertical shots, provokes the distortion of the shapes, and therefore, their perception becomes more complex and challenging. On the other hand, as it happens with an abstract design, taking distance from the surface

[10]I would like to point out that there are other possible outcomes of colour digital edition even if they are not part of Dronestragram visual universe. For example, the artist Florian Maier-Aichen's edition of the colour consists precisely of leaving traces of the digital intervention. In other words, this artist makes visible the digital edition on the visual composition of the image.

[11]First prize winner, category Nature 2015 Contest: Snorkelling with sharks. Photographer Tahitiflyshoot.

[12]Although I am referring particularly to the nonlinear patterns of natural vegetation's distribution in comparison to the linear patterns that cultivated fields have, I would like to acknowledge the symmetries hidden in nature as well as the dynamism, instability and complexity of its behaviour (Prigogine and Stengers 2004).

makes invisible some details and makes visible patterns: the patterns of colours, the patterns of shapes, the patterns of textures. These two changes provoke the viewers need to keep on looking in order to decipher what the photograph is showing. In other words, drone photography shows its audiences a nature that they already know and that they have already seen but in a new way. Whilst the vertical or oblique proximity perspective makes the image memorable through the changes on the perception of natures' shapes, the digital production of the colour gives clues to identify shapes and patterns, as well as to reinforce the perception of *the real* through the image. Either through the shapes or through the colours, the vertical or oblique proximity perspective contributes to show the beauty of highly produced natures.

6.6 The Circulation of Drone Photography or Its Co-production

The 'most amazing aerial view' of the 2014 contest, was awarded to a drone photography that shows an eagle in flight which seems to be looking at both the drone and the landscape of the Bali National Park (Indonesia).[13] The photographer explains:

> The photo was taken on 2012, we were working on aerial photos for eco-tourism spot at Bali Barat National Park (west Bali, Indonesia). We were taking vertical aerial photos using autopilot. Suddenly the eagle came and chasing our drones, so I decided to take over and switch to manual control. After few minutes we discovered that the eagle did not attack the drone but chasing and playing with it. So we decided to do manual flying (with fpv) and make the eagle chase the drone and do manoeuvre so the eagle positioned just in front our camera. So it just a huge luck and a perfect moment combination.[14]

The picture was taken in a national park, insinuating either nature appreciation or eco-tourism, and it is appealing to a global audience engaged in a wide range of environmental discourses. As Cosgrove and Fox (2010: 87) state 'in the popular imagination, aerial images of the earth's surface and landscapes have come to be framed almost exclusively through an environmental lens'. This popular imagination arises more complex the tensions, as some of the users' comments reveal:

> Reckless and irresponsible photography wins a prize. Great.
> No explanation is good enough to defend the risk of manoeuvring a drone near a bird (or any other wild life for that matter). What would the consequences be if there was a collision? This is irresponsible and should be treated as a criminal offence! Stiam Jakobsen[15]

Comments like these openly put into question the environmental engagement of this photo, as well as the ethics of the photographic act. Whilst Dronestagram users recognise the image's beauty, they are also pointing out the responsibilities that the production of a drone photography entails.

The main mode to exhibit images in Dronestagram, whether still images or videos, is as a mural or as the composition of images indexed by categories or tags. The murals of drone images, in general, have three or six columns and almost infinite number of rows. Edward Tufte (2012: 105) explains that 'multiple images reveal repetition and change, pattern and surprise' and furthermore 'amplify, intensify, and reinforce the meaning of images'. The adjacent exhibition of multiple images helps to detect contrasts, correspondences, analogies and patterns. If categories and tags shape the interpretation of the drone images, the murals constructed by Dronestagram produce visual subjects to make sense of those categories and tags. Beyond the meaning imagined by the photographers, I will suggest that Dronestagram is not only a site to publish and to view the drone photographs, but it is also involved in the production of modes of seeing them. In effect, through the composition of groups of images, Dronestagram presents an

[13]First prize winning photography 2014 contest. Photographer: Capungaero.

[14]https://diydrones.com/profiles/blogs/first-international-drone-photography-awards-and-the-winner-is.

[15]See this and other comments at http://www.dronestagr.am/bali-barat-national-park-indonesia/.

organisation of the visual universe of drone photography producing meaning and shaping the modes of seeing.

The contests are also another type of montage widely used in Dronestragram that make users associate the drone images to narratives and discourses. The contest 'Small drones, big changes', to take just an example, was launched one month earlier to the 21st Session of the Conference of the Parties to the United Nations Framework Convention on Climate Change (Paris, 2015). Besides the slogan of the contests, the winning photographs exhibit consists of visual pairs that present *problem/solution*. The visual parallelism is what makes visible the connection between a particular environmental problem and a possible way to tackle it. Hence, this contest forces the viewers to interpret each photo of the visual pairs within an environmental discourse. Without this context, some of the photographs could perfectly fit in the category of Nature (and in fact the user can find them in this category) or become the cover of an architecture, industrial design or engineering magazine.

6.7 Concluding Remarks: The Politics of the Drone Images

Harun Farocki and Kaja Silverman focus on one scene in Godard's film *Le Gai Savoir* (1968) in which Patricia, one of the main characters of the film asks 'What is a false image?' and she answers, 'There, where the image and sound [seem] true'. Farocki and Silverman point out: 'She thereby suggests that an image or sound is false not when it misrepresents reality, but when it seems adequate to reality—when we are able to assimilate it to ourselves. An image or sound is true, on the other hand, when it manifestly fails to represent us' (Farocki and Silverman 2016: 187). By looking at the visual universe of Dronestragram, in this chapter, I tried to argue that we need to put drone images into question within the civil realm too.

The hyperrealism of drone images is simultaneously the base upon which they are considered faithful images, and what precludes to identify and analyse all the things that these images are leaving out. The politics of drone images operates precisely through a composition of silences and absences. Let me give an example to demonstrate the politics of these images within the civil realm. Last year, just a few hours after a young man of a *Mapuche*[16] community was murdered in the southern Argentinean province of Rio Negro, the newspaper *La Nación* published on its online edition an article entitled 'Villa Mascardi from a drone: This is the tourist area that today concentrates the *Mapuche* conflict.' The visual part of the article was a drone video. In less than two minutes the video was showing a smooth flight over a paradise in Patagonia: the lake, the mountains and the forest were beautiful and captivating. The proximity perspective (combining vertical and oblique images) contributed to make the readers believe that the drone view did not leave anything out of their eyes. However, neither *Mapuches* people, who were claiming the right over an ancestral territory, nor the police, who were involved in the murder, were part of the drone video. I should stress that although the video had the characteristics of a Patagonia's touristic destination advertisement, the article's visual narrative presented only the tip of the iceberg of a conflict of land appropriation in Patagonia. Instead of acknowledging the long history of *Mapuches* land's rights, the drone video puts forward the idea that indigenous communities are not part of that territory. I think that the editors' choice to include a drone video as part of that article is directly related to the status that drone images have acquired in the popular visual culture: people believe they can find the truth through these images.

Analysing how the aura of transparency of this genre of photography is constructing nature may allow us to contemplate both the regimes of visibility created through this technology and the politics of the drone images within the civil realm. As Doreen Massey has put it: 'Not all views from above are problematical. The

[16]The Mapuche people are the original inhabitants of the Southern territories of Argentina and Chile.

problem only comes if you fall into thinking that vertical distance lends you the truth' (Massey 2007: 107). In other words, we need to analyse how we look at drone images in civil practices, too.

I have also argued that the montages of the drone photographs produce the image and its meaning. The production of meaning through the montage, hence, may let us think that we can construct other possible montages for the drone images. I think that as geographers we have the responsibility and the challenge to imagine other possible montages aiming to put into question our ways of looking at these awesome images, as well as our position towards them. We need to make other compositions in order to start acknowledging all the silences that these amazing images have.

References

Arnheim R (1997) Visual thinking. University of California Press, Berkeley, Los Angeles and London

Ball P (2013) La invención del color. Fondo de Cultura Económica, Madrid

Birtchnell T, Gibson C (2015) Less talk more drone: Social research with UAVs. J Geogr High Educ 39 (1):182–189

Bridle J (2014) The Drones among us. https://www.politico.com/magazine/gallery/2014/02/the-drones-among-us-183694?slide=11. Last access 24 Sept 2018

Chamoyou G (2015) A theory of the Drone. The New Press, New York

Cosgrove D, Fox W (2010) Photography and Flight. Reaktion Books, London

Farocki H, Silverman K (2016) A propósito de Godard. Caja Negra, Buenos Aires

Garrett B, Anderson K (2018) Drone methodologies. Taking flight in human and physical geography. Trans Inst Br Geogr 43:341–359

Graham S (2016) Vertical: the city from satellites to bunkers. Verso, London

Gregory D (2010) Seeing red: Baghdad and the event-ful city. Political Geogr 29:266–279

Gregory D (2011) From a view to a kill: Drones and late modern war. Theory, Cult Soc 28(7):188–215

Gregory D (2015) Moving targets and violent geographies. In: Merrill H, Hoffman L (eds) Spaces of danger: culture and power in the everyday. University of Georgia Press, Athens, pp 256–296

Hawkins S (2010) American iconographic: national geographic, global culture and the visual imagination. University of Virginia Press, Charlottesville

Kindervater KH (2015) Lethal surveillance. Drones and the geo-history of modern war. Phd Dissertation. University of Minnesota, Minneapolis

Massey D (2007) For space. Sage, London

Mirzoeff N (1999) Una introducción a la cultura visual. Paidós, Buenos Aires

Prigogine I, Stengers I (2004) La nueva alianza. Metamorfosis de la ciencia, Alianza Editorial, Madrid

Rose G (2016) Rethinking the geographies of cultural 'objects' through digital technologies: interface, network and friction. Progr Human Geogr 40(3):334–351

Rothstein A (2015) Drone. Bloomsbury Publishing, New York

Ryan J (2011) Travels through the Magic Lantern. In: Fiell C, Ryan JR (eds) Memories of a lost world. Travels through the magic lantern. Fiell, London, pp 16–30

Tufte E (2012) Visual explanations: images and quantities. Evidence and narrative. Graphic Press, Cheshire

7 "Our Field Is the World": Geographical Societies in International Comparison, 1821–1914

Maximilian Georg and Ute Wardenga

Abstract

As associations for the promotion and dissemination of geographical knowledge, Geographical Societies were the institutional basis of geography for the larger part of the "long" nineteenth century. Before 1914, up to 170 such Societies existed in all inhabited continents. Most historiographical research has focused on Geographical Societies in capital cities and/or dealt with them as being inside the "containers" of their respective nation-states, and as if they existed and operated in independence and isolation from one another. By contrast, in a research project launched in 2015/16 at the Leibniz Institute for Regional Geography (IfL), Leipzig, within the framework of the Leipzig Collaborative Research Centre (SFB) 1199 "Processes of Spatialization under the Global Condition", we seek to identify connections and draw comparisons among 34 Geographical Societies from all continents and of 13 languages, including Societies from minor cities and countries. Our data comes from the Societies' yearly journals, which we record with a standardized method that we have developed. From a Society's proceedings, we gather, rather qualitatively, its organizational structure including its networking with other Geographical Societies; from the geographical articles in its journal, we gather, rather quantitatively through codes, the Society's subjects (e.g. "physical geography": "geology"; or "human geography": "economy"), and world areas (e.g. "Africa, West", or "Polar Regions, South") of interest. For each Society, we thus obtain a profile reflecting its structure, activities, interests and evolution. Each profile may be explained by the Society's local and historical context (e.g. French colonialism; Czech nationalism), and further understood through the theoretical concepts of our Collaborative Research Centre: each Society "spatialized" the world into certain "spatial formats", which then made up a certain "spatial order". By negotiating modes of dealing with a globalized world, the Geographical Societies thus contributed to the professionalization of geography.

Résumé

En tant qu'associations pour la promotion et la diffusion de savoir géographique, les Sociétés géographiques furent le fondement institutionnel de la géographie pour la plupart du «long» XIXe siècle. Avant 1914, il existait jusqu'à 170

M. Georg · U. Wardenga (✉)
Leibniz Institute for Regional Geography, Leipzig, Germany
e-mail: U_Wardenga@leibniz-ifl.de

M. Georg
e-mail: M_Georg@leibniz-ifl.de

B. Schelhaas et al. (eds.), *Decolonising and Internationalising Geography*, Historical Geography and Geosciences, https://doi.org/10.1007/978-3-030-49516-9_7

Sociétés géographiques dans tous les continents habités. La plupart de la recherche historiographique s'est concentrée sur les Sociétés géographiques des capitales et/ou les a regardées à l'intérieur des «containers» de leurs États-nations respectifs, et comme si elles existaient et fonctionnaient de manière indépendante et isolée les unes des autres. En revanche, dans un projet de recherche entamé en 2015/16 à l'Institut Leibniz de géographie régionale (IfL), Leipzig, dans le cadre du Centre de recherche collaborative (SFB) 1199 «Processus de spatialisation à l'époque de la mondialisation», Leipzig, nous voulons mettre en évidence des relations et faire des comparaisons entre 34 Sociétés géographiques de tous les continents et de 13 langues, y compris des Sociétés de villes et de pays mineurs. Nos données viennent des journaux annuels des Sociétés, que nous enregistrons par une méthode standardisée que nous avons développée. Des actes d'une Société, nous recueillons, d'une manière plutôt qualitative, sa structure organisationnelle, y compris ses relations avec d'autres Sociétés géographiques. Des articles géographiques dans son journal, nous recueillons, d'une manière plutôt quantitative par le biais de codes, les sujets (par exemple, «géographie physique»: «géologie», ou «géographie humaine»: «économie») et régions du monde (par exemple, «Afrique, Ouest», ou «Régions polaires, Sud») qui intéressèrent la Société. Pour chaque Société, nous obtenons ainsi un profil reflétant sa structure, ses activités, ses intérêts et son évolution. Chaque profil peut être expliqué par le contexte local et historique de la Société (par exemple, le colonialisme français ou le nationalisme tchèque), et être compris davantage par le biais des concepts théoriques de notre Centre de recherche collaborative: chaque Société «spatialisa» le monde en certains «formats spatiaux», qui, eux, constituèrent un certain «ordre spatial». En négociant des modes de gérer un monde mondialisé, les Sociétés géographiques contribuèrent ainsi à la professionnalisation de la géographie.

Resumen

Como asociaciones para la promoción y difusión del conocimiento geográfico, las Sociedades Geográficas fueron la base institucional de la geografía para la mayor parte del "largo" siglo XIX. Antes de 1914, existían hasta 170 sociedades de este tipo en todos los continentes habitados. La mayor parte de la investigación historiográfica se ha centrado en las Sociedades Geográficas en las ciudades capitales y/o las ha tratado como si estuvieran dentro de los "contenedores" de sus respectivos Estados nacionales, y como si existieran y operaran en independencia y aislamiento la una de la otra. Por el contrario, en un proyecto de investigación lanzado en 2015/2016 en el Instituto Leibniz de Geografía Regional (IfL), Leipzig, en el marco del Centro de Investigación Colaborativa de Leipzig (SFB) 1199 "Procesos de espacialización bajo la Condición Global", buscamos identificar conexiones y establecer comparaciones entre 34 Sociedades Geográficas de todos los continentes y de 13 idiomas, incluidas las Sociedades de ciudades y países menores. Nuestros datos provienen de las revistas anuales de las Sociedades, que registramos con un método estandarizado que hemos desarrollado. A partir de los procedimientos de una sociedad, reunimos, de forma bastante cualitativa, su estructura organizativa, incluida su creación de redes con otras sociedades geográficas. A partir de los artículos geográficos en su revista, reunimos, cuantitativamente, a través de códigos, los temas de la Sociedad (por ejemplo, "geografía física": "geología"; o "geografía humana": "economía") y áreas mundiales (por ejemplo, "África, Occidente", o "Regiones polares, Sur") de interés. Para cada Sociedad, obtenemos un perfil que refleja su estructura, actividades, intereses y evolución. Cada perfil puede explicarse por el contexto local e histórico de la Sociedad (por ejemplo, el colonialismo francés; el nacionalismo checo), y entenderse mejor a través de los conceptos teóricos de nuestro Centro de Investigación Colaborativa: cada Sociedad "espacializó" el mundo en ciertos "formatos espaciales", que luego formó un cierto "orden espacial". Al negociar modos de tratar con un mundo globalizado, las Sociedades Geográficas contribuyeron así a la profesionalización de la geografía.

Keywords

Geographical societies · History of scientific institutions · History of geography · Spatialization · Global condition · Geographical journals · Coding · Transnational history

7.1 Introduction

The title of this paper is the title of a Leipzig research project (funded by the German Research Foundation—DFG) of which U.W. is the originator and leader, and M.G. one of the main researchers. At the 25th International Congress of History of Science and Technology in July 2017, M.G. outlined the project, which was launched in late 2015/early 2016. In the following, we represent that outline and substantiate it with the additional insights we have gained between the Rio de Janeiro Congress and November 2019, as we are finishing this paper. In Sect. 7.2, we define our subject, Geographical Societies, and explain the shortcomings of previous research on them that our project seeks to rectify. In Sect. 7.3, we present our project's aims and methods, and in Sect. 7.4 its preliminary results.

7.2 What are Geographical Societies?

As associations for the promotion and dissemination of geographical knowledge, Geographical Societies were the institutional basis of geography for the larger part of the "long" nineteenth century. They organized or supported expeditions to unknown world regions, and presented the results of that and other geographical research to a wide public, in their publications, talks and exhibitions. They organized national or international geographical congresses, and worked to establish geography as a subject both in schools and universities. Thus, it was mainly thanks to Geographical Societies that geography transformed into a professional scientific discipline.

The first Geographical Societies emerged in Paris (1821), Berlin (1828) and London (1830). Before 1870, other European and American political or commercial centres followed suit: Frankfurt/Main (1836), St. Petersburg (1845), Mexico City (1851), New York (1851), Vienna (1856), Geneva (1858), Leipzig (1861), Dresden (1863), Rome (1867; until 1871/72 based in Florence), Munich (1869). The "fashion" reached its peak in the 1870s and 1880s, as each year brought up to 10 new Societies, now also in capitals of minor European countries (Budapest 1872, Amsterdam 1873, Bern 1873, Bucharest 1875, Lisbon 1875, Brussels 1876, Copenhagen 1876, Stockholm 1877, Edinburgh 1884, Christiania 1889); in minor cities of France (e.g. Bordeaux 1874, Montpellier 1878, Lille 1882), of Germany (e.g. Halle/Saale 1873, Bremen 1877, Cologne 1887), of Switzerland (e.g. St. Gall 1878, Neuchatel 1885) and of England (Manchester 1884, Newcastle/Tyne 1887); in capitals of North Africa (Cairo 1875, Algiers 1879),[1] of Asia (Tokyo 1879), of the Australian colonies (e.g. Melbourne 1883, Brisbane 1885) and of Latin America (Rio de Janeiro 1883, Lima 1888, La Paz 1889). Between the 1890s and 1914, Societies were founded less frequently, but overall still in great number, and in an even greater variety of cities, including several in the U.S. (e.g. San Francisco/California 1891, Philadelphia 1892, Chicago 1898).

In total, before 1914, up to 170 Geographical Societies plus dozens of local branches of theirs existed at least temporarily—some Societies did not last long—in all inhabited continents

[1] At the time, to be sure, Algeria was a French colony, and Egypt was about to become a British one. Thus, their Geographical Societies consisted of immigrated Westerners and Westernized locals rather than of "real" Egyptians or Algerians; and the language of both the Cairo and the Algiers Societies was French rather than Arabic.

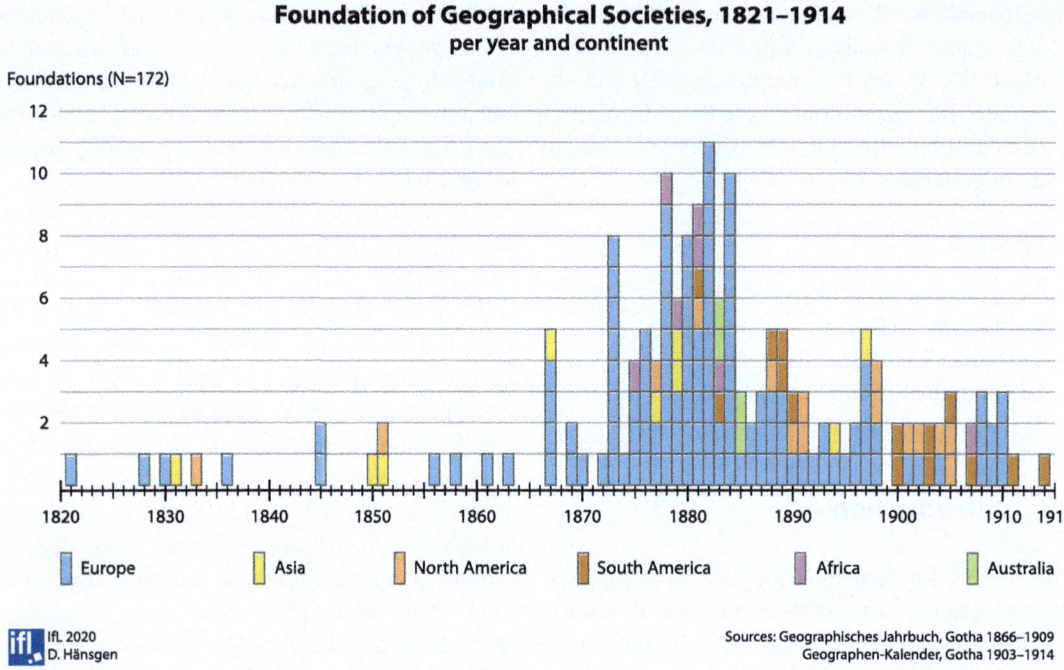

Fig. 7.1 Foundation of Geographical Societies, 1821–1914

(Fig. 7.1).[2] They all had features in common: they dealt with geography, were headed by a president, invited speakers, awarded decorations, published periodicals. On the other hand, they differed in terms of their countries and the significance of their cities, in their financial and infrastructural means, in the number and background of their members, in their relations with their national governments, in their degrees of professionalism, and in their regions and subjects of interest.

For example, the Societies of Paris and London, respectively, were based in capitals of global empires, whereas Leipzig was the seat of no government, and Saint-Nazaire (est. 1886) was a town rather than a city. The Societies of Saint-Nazaire and Stuttgart (1882) called themselves "Societies of Commercial Geography", whereas most other "Geographical Societies" did not delimit their subjects of interest in their names.[3] In terms of resources, the Societies of London and St. Petersburg had plenty of money, while those of Halle and Brisbane had very little at their disposal. The Societies of London and Washington/D.C. (1888) possessed thousands of members, those of Bordeaux and Madrid (1876) some hundreds, and those of Cologne and La Paz only some dozens (statistics for 1908, in: Kollm 1909: 411–418).

[2]The most-quoted lists of Geographical Societies are those compiled by, successively, Ernst Behm, Hugo Wichmann and Georg Kollm in various volumes of the *Geographisches Jahrbuch* between 1866 (vol. 1) and 1909 (vol. 32). However, for the time after 1909, and because of omissions and errors in those lists, we have complemented them with various other sources. By the way, in contrast to other research, we do not count as Geographical Societies those establishments, mostly located in Latin America, that called themselves "Geographical Institutes", such as the "Institutes" of Rio de Janeiro (est. 1838) and Buenos Aires (1879).

[3]The name "Society of Commercial Geography" meant that the respective Society was led by businessmen and other "bourgeois" and sought to collect geographical information that would help overseas businesses and trade, especially in or with the respective country's colonies. In regions and subjects of interest, Societies of "Commercial Geography" did not differ much from "classic" Societies of Geography. The difference lay in methods and aims: "commercial geography" was an applied, "utilitarian" one, as opposed to a theoretical, academic geography (Lejeune 1993: 147–156).

7.2.1 Previous Research on Geographical Societies and Its Shortcomings

So far, two types of histories of Geographical Societies have been published, one acritical and the other one critical.

First, Societies have been celebrated on the occasion of their anniversaries—for instance, 25 years of Hamburg (Friederichsen 1898); 50 years of Berlin (Koner 1878), London (Markham 1881) and Brussels (Société royale belge de géographie 1926); 75 years of Washington (Payne 1963); 80 years of Quebec (Camu 1957); 100 years of Copenhagen (Schou 1977), Toulouse (Camboulives 1981/82) and Zurich (Jud 1989); 120 years of Budapest (Marosi 1994); 150 years of Vienna (Kretschmer and Fasching 2006). Those congratulatory chronicles started to emerge as early as the second half of the nineteenth century. Usually, they are penned, or delivered before the Societies, by leading members of theirs, and published by the Societies themselves.

Second, historians of geography have revealed nationalist, colonialist and/or imperialist dimensions of the activities of Geographical Societies.[4] In fact, those associations, and the very discipline of geography,[5] did not collect "innocent" data on the Earth, but promoted and disseminated—voluntarily or not—knowledge that could and did enable governments and private actors to exercise power over national and foreign territories.

Despite the large number of historiographical publications on Geographical Societies that we have cited above, that research suffers from three significant shortcomings, of which the first contributes to the second, and the first and second jointly contribute to the third.

First, the Geographical Societies of capital cities have received much more attention than those of minor ones. Histories of both types defined here have been written on most of the capital Societies, with some—especially on London—even zeroing in on specific issues such as the Royal Geographical Society's gender policies (Bell and McEwan 1996) and professionalization (Hayes 2018). On most "provincial" Societies, by contrast, there are either "Festschriften" or no histories at all, although there have been signs of improvement (in the form of Péhaut 1994; MacKenzie 1995; Bonnefont 1999; Clout 2008a, b; Barke et al. 2013).

Second, in most cases, the Geographical Society of a capital city also is *the* national or royal/imperial Geographical Society of its country. In fact, in many countries, the capital has been the sole city with a Geographical Society. However, by focusing on capital Societies, histories of Geographical Societies have largely remained within the "containers" of nation-states (perhaps including their empires); of national geographies and national publics. For instance, the "Royal Geographical Society" of Madrid has been studied with respect to "Spanish colonialism in Morocco" (Villanova Valero 1999), the "Imperial Russian Geographical Society" of St. Petersburg as the originator of "Russian ethnography" (Berelowitch 1990), the "National Geographic Society" of Washington as an influencer of "the geographical imagination in America" (Schulten 2001: Chaps. 3, 7).

Third, as a consequence, few histories of Geographical Societies cover more than one Society, and if they do, it is again within the borders of France (Lejeune 1993), Spain (Hernández Sandoica 1994), Belgium (Nicolaï 1994),

[4]On the Societies of Rome: Carazzi (1972), Natili (2008); Milan: Milanini Kemény (1973); Berlin: Bader (1978); St. Petersburg: Bassin (1983), Weiss (2007); Bordeaux: Péhaut (1994); Lima: López-Ocón (1994); Quebec: Bergevin (1994); Amsterdam: van der Velde (1995); Cairo: Ferrié and Boëstch (1996); Madrid: Rodríguez Esteban (1996), Villanova Valero (1999), Nogué and Villanova (2002); Nancy: Bonnefont (1999); London: Driver (2001): Chap. 2; Edinburgh: Kuitenbrouwer (2004); Lisbon: Garcia (2004); Paris: Gómez Mendoza (2005); Lyon: Klein (2008); France: Lejeune (1993); Spain: Hernández Sandoica (1994); Belgium: Nicolaï (1994); Italy: Monina (2002); Latin America: López-Ocón (1996); on various Geographical Societies: Capel (1981): Chap. 7, Heffernan (2003), Butlin (2009): Chap. 6.

[5]On geography's links to "national identity": Hooson (1994); to "colonization": Bruneau and Dory (1994), Singaravelou (2008); to "imperialism": Hudson (1972), Bell et al. (1995); to "empire": Godlewska and Smith (1994), Driver (2001), Butlin (2009).

Italy (Monina 2002), or those and other countries in juxtaposition (Butlin 2009). Yet where are the transnational connections among those 170 Geographical Societies that existed in the world before 1914, and where are the transnational comparisons among them—as their connections would be shaped by their similarities and differences?

In fact, the world's Geographical Societies observed, and reported on, each other's activities. They exchanged publications and letters. They awarded medals or corresponding or honorary memberships to other Societies' members, or invited them as speakers. They met at the national and international geographical congresses that they organized. They sought to emulate other Societies out of admiration, or to outstrip them out of national or imperial-colonial rivalry. Thus, for example, the "Imperial Russian Geographical Society" of St. Petersburg may have acted as an "international representative of the Russian empire" (Weiss 2007: Chap. 6) also towards Geographical Societies outside of Russia, and it makes a lot of sense to connect and compare the Societies of two countries such as Argentina and Brazil (Zusman 1996); of three such as France, Britain and Prussia (Péaud 2016: 79ff.); of an entire continent such as Latin America (López-Ocón 1996); or of the entire globe (Capel 1981: Chap. 7; Heffernan 2003). Nevertheless, the "transnational" histories of Geographical Societies cited here have so far remained exceptions.

7.3 Our Leipzig Research Project

With this project, we attempt to compose such a "transnational" history of Geographical Societies. We want to understand them as a global phenomenon without methodological nationalism. We want to make them and their structures, activities and interests comparable both synchronically and diachronically. We want to demonstrate both explicit and implicit entanglements among them. The project is based at the Leibniz Institute for Regional Geography (IfL) in Leipzig and belongs, under number C01, to the Leipzig Collaborative Research Centre (SFB) 1199 "Processes of Spatialization under the Global Condition" (funded by the German Research Foundation—DFG). Within the Centre's theoretical framework (Middell 2019), we ask how Geographical Societies contributed to new "spatial formats" and new "spatial orders" in the world, such as nation-states and empires.

Since the project's beginning in 2015/16, we have looked at the "long" nineteenth century and at a sample of 34 Geographical Societies. We reached that number first by selecting the Societies that called themselves, in their respective languages, "Geographical Societies" (as opposed to "Societies of Commercial Geography"), and that published a yearly periodical from not much later than 1890 onwards. From that pre-selection, we then selected those Societies whose periodicals would be available to us in libraries near Leipzig or online, and written in languages that ourselves or student assistants would be able to read. In total, our sample covers Geographical Societies from all inhabited continents and of 13 languages, allowing us to identify connections and draw comparisons on a truly global scale.

7.3.1 Method of Data Collection

Our primary sources are the Societies' journals, since a sample as large as ours would not allow us to base our research on archives. The journals contain plenty of information: geographical treatises, travelogues, geographical news, reviews of publications, obituaries, and the respective Society's proceedings and correspondence. In order to compare the Societies' profiles on that basis, we have developed a standardized method of recording their journals (Steinbach-Hüther et al. 2019: 12–19). First, for each volume, we record, in a Microsoft Word file, what it contains on the respective Society's structure: foundation, by-laws, leadership, membership, infrastructure, characteristics of the journal, etc. Most of that information comes from the Societies' proceedings. Second, for each journal, we record, in a Microsoft Excel file, each volume's geographical articles with their authors and bibliographical data (volume number, page range, etc.). Moreover, for each article, we define, by means of codes, its external and internal types as well as the subject and world area it addresses.

An article's external type, recorded in a first Excel column, may be "treatise", "reprinted document/list", or "obituary"; its internal one, recorded in a second column, "research paper", "expedition", "exchange of publications", etc. Where we assign both the codes "reprinted document/list" and "exchange of publications", Geographical Societies list, in their journals, other Geographical Societies with which they were exchanging publications at the time—one of the connections among Societies to which our research project pays special attention. In a third Excel column, we record an article's general subject, distinguishing among "physical geography", "human geography", "cartography" or, in mixed cases, "general geography". As particular subjects, we distinguish, in a fourth column, "geology", "wildlife", "ethnic groups", "economy", etc.

For areas, recorded in a fifth Excel column, we have divided the globe into 23 parts. There is no perfectly "natural" regionalization of the world, because regionalization can follow many different criteria. For example, does a sea separate or, on the contrary, connect its coasts? At least, in our regionalization, we have tried to overcome traditionally imagined spaces such as "the Orient" and "Eastern Europe". Therefore, we have assembled our 23 world areas out of the sheets of the "Karta Mira", for in this 1:2,500,000 scale "World Map", created from 1960 to 1980 by the Soviet Union, the GDR and five other socialist states, sheet lines simply run along regular intervals of latitudes and longitudes. Along a practical selection of these lines, we have defined, as codes, areas such as "Indian Ocean, North" (14), "South America, West" (17), and "Polar Regions, South" (23) (Fig. 7.2). In addition, there are 11 combinations of areas 01–23, such as "Europe" (26; consisting of 03—Europe, North, and 08—Europe, South), and "Pacific Ocean" (28; consisting of 05—Pacific Ocean, North, and 11—Pacific Ocean, South). Finally, there are combinations of all 23 areas, coded as "Global" (35), and of single areas distant from each other, coded as "Transregional" (36).

Recording our sample of 34 Geographical Societies with this method has so far resulted in a total of about 1,000 Word files (i.e. journal years/volumes) and 60,000 Excel lines. In the Excel spreadsheet of each Society, we may freely search for any keyword or filter the lines for it, and especially for the codes in the five coded columns. For example, we could find all "research papers" in a journal that deal with "geology" of world area "Asia, East" (10); or all reports of "expeditions" published between 1881 and 1905 that travelled to "Africa, West" (19); or all years in which the Geographical Society of Bremen exchanged, according to its journal, publications with the Geographical Society of Lisbon; and so on and so forth. For any article type, author, subject, or area or combination thereof, a Society's Excel file allows us to instantly find how the keyword(s) or code(s) appear(s) in the Society's journal, and how the instances have evolved in the course of the journal's life.

7.3.2 Working Papers and Publications

For each fully recorded Geographical Society, we use the Word and Excel files to portray its structure, activities, interests and evolution in 5,000 to 20,000 words, depending on the size of the data. In 2020/21, we will publish the portraits in a volume, including a portrait of the Geographical Society of Tokyo, although for organizational reasons, the main author, Ling Chen (in 2018: Leipzig University, Chinese Studies), has analyzed its journal cursorily rather than with our above-described method.

Also in 2020/21, we will publish a volume with papers on specific research questions related to various Geographical Societies in the "long" nineteenth century. Some of them were presented in a panel that we organized at the 5th European Congress on World and Global History in Budapest in September 2017. In the volume,

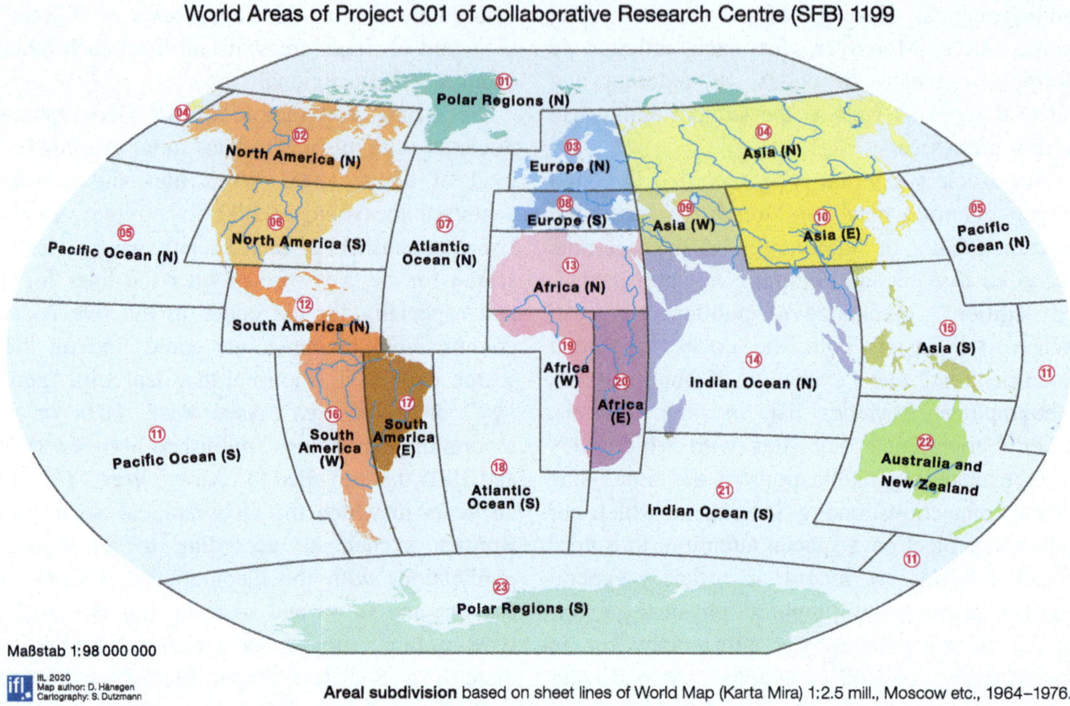

Fig. 7.2 World areas for coding regional interests of Geographical Societies

members of our project as well as external historians of geography will discuss Geographical Societies of one or several countries, to make the comparison also transnational. M.G. analyzes how India was seen in the journals of the Societies of Brussels, Edinburgh and Washington; and he and Maximilian Stintzing analyze the visits of Henry Morton Stanley, explorer of Africa, to various Geographical Societies. In three other contributions, the authors will summarize their respective 2017 Master's thesis, which they wrote in association with our project, using our method of data collection: Tomas Cerny (in 2017: Leipzig University, Global Studies) on the Geographical Society of Prague; Paul Greiner (in 2017: Roskilde University, Global Studies) on the Societies of Copenhagen, Stockholm and Christiania (Oslo); and Maximilian Stintzing (in 2017: Bamberg University, Geography) on the Societies of Lima and Madrid. Another research assistant of ours, Vincent Schober (in 2018: Leipzig University, Geography), will not appear in the conference volume, but has written, likewise with our method, a 2018 Bachelor's thesis on the Geographical Society of Amsterdam.

7.4 Preliminary Research Results

The Geographical Societies' journals have turned out to be more heterogeneous than one would have expected. They vary not only in the number of pages per year (from over 100 to over 1000), but also in structure and contents. For example, most articles of the Marseille Society report on travels and are shorter than 10 pages, compared to which the Nancy Society prefers geographical treatises of a few dozen pages. Or, not all journals necessarily contain a separate section for reviews of geographical publications. Also, large Societies such as London and Rome have at some point published their proceedings in volumes separate from those containing geographical articles.

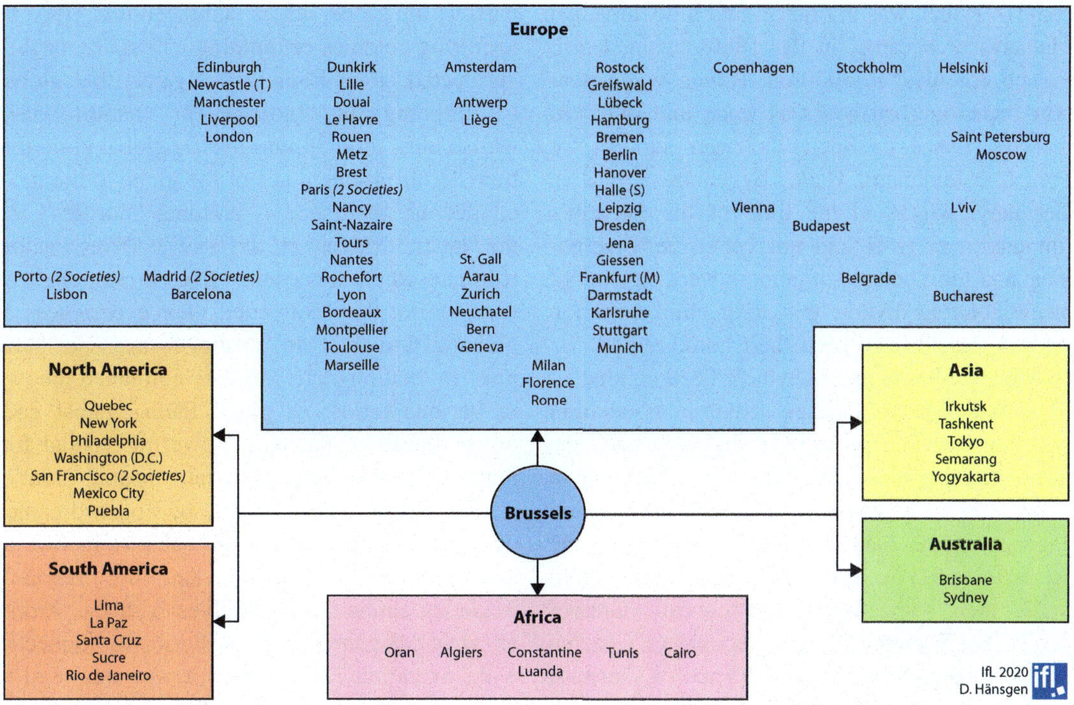

Fig. 7.3 Geographical Societies with which the Geographical Society of Brussels exchanged journals and other publications, 1876–1914

Another example: each volume of the Brussels Society lists the latter's exchange partners, whereas the journal of the Washington Society never contains such lists. However, Brussels states that they exchanged, every year from 1889 to 1914, publications also with Washington. The Washington journal may remain silent on this because its editors did not have time for the extra work, or did not deem it necessary; or because they listed the exchange partners outside of the journal.[6] On the other hand, Brussels's meticulous publication of exchange partners as well as the latter's sheer number illustrate how important networking with other Geographical Societies (and other learned institutions, which are likewise listed) was for the Belgians' self-concept. From its foundation in 1876 until 1914, the Brussels Society exchanged publications with, in total, 95 Geographical Societies from all continents, which accounts for around 60% of all Societies existing during that period (Fig. 7.3).

Finally, examples of different subjects and areas of interest: the Society of Philadelphia dealt with physical geography, in particular geology, more than with human geography, whereas the Society of Manchester dealt with human geography, in particular communications, trade, history and ethnic groups, more than with physical geography. Or, the Society of Halle dealt mostly with the area "Europe, North" (03), in particular Halle's immediate (German) surroundings, whereas the Society of Rouen dealt, no less than with its own area (03), with that of "Indian Ocean, North" (14), in particular Indochina and Madagascar.

The next step is to explain *why* the Geographical Societies had the specific interests that their journals reveal. Answers are first derived from a Society's structure and local and historical

[6]In the case of, for example, the Rome Society, its journal's monthly issues were each delivered with a cover containing, among other things, a list of the Society's current officials. At libraries, however, all issues of one year were usually bound together in one volume, in which process the issues' covers were discarded. Therefore, the information on the covers may elude us, because for our research project, we must do with the printed or scanned library copies of journals that we happen to obtain.

context. Rouen was in France, and Indochina and Madagascar became, in the 1880s at the latest, French colonies. In addition, Rouen was a seaport depending on overseas trade and overseas markets, hence its Society's eager support for French colonialism. Halle, by contrast, was in Germany, which, at the time of the Society's foundation in 1877, did not yet strive for colonies; and the Society's members were citizens of professions so diverse that their chief common interest should have been their home region.

More profound analyses have been conducted in the three Master's theses written in association with our research project. In their journals, the three Geographical Societies of Scandinavia (Copenhagen, Stockholm and Christiania) were eager to study, map and survey regions of the Arctic (area 01) because they hoped to bring them under their respective state's control (Greiner 2017). The Society of Prague focused on Czechia and its surroundings (in areas 03 and 08) because it wished to emancipate the Czech nation from the suzerainty of the Habsburg Empire (Cerny 2017). Moreover, the Prague as well as the Rouen Societies consisted mainly of schoolteachers. Inspired by their role as servants of their states and educators of their people's next generation, those teachers may have pressed ahead with Prague's nationalism as well as Rouen's colonialism.[7] The Society of Lima, in turn, consisted of Peruvian political, military and economic elites. Accordingly, its journal dealt almost exclusively with Peru, because the elites of that country—hard to access and yet to be consolidated—were seeking to territorialize their rule over it, and to exploit its natural resources for export to Europe and the U.S.A. (Stintzing 2017).

Second, we may understand the interests and activities of Geographical Societies through the theoretical concepts of our Collaborative Research Centre "Processes of Spatialization under the Global Condition". In reaction to the increasing interconnectedness of the different parts of the globe, actors embraced that globe by acquiring colonies or building an empire, or they "protected" their homeland "against that globe" by building an "independent" nation-state—whose idea, though, was itself a global good that travelled from one place of the globe to the next, connecting them across national "borders" (on the latter: Meyer et al. 1997). Thus, actors restructured the spaces they lived in, respatialized their world: they converted their experiences of globalization into the "spatial formats" of "empire" or "nation-state"; or they combined the two in various ratios, or pitted them against each other. Together, the most powerful "spatial formats" then constituted a certain "spatial order".

The Geographical Society of Rouen dreamed of a global empire of French colonies in various continents. The Scandinavian, or "Nordic", Societies contented themselves with the Arctic, an area adjacent to, and historically connected with, Scandinavia, of which it was supposed to be an imperial "extension" (German: *imperialer Ergänzungsraum*). The Societies of Lima and Prague each opted for the nation-state, but as Prague did not yet belong to an independent country, it had to pit its nation-state against the "empire format" of Austria-Hungary—and, we assume, of the Geographical Society of Vienna, the imperial capital. The Society of Madrid, by contrast, pitted empire against nation-state: between the beginning and the end of the nineteenth century, Spain lost almost all its colonial possessions, which once had constituted the largest empire in the world. However, the Madrid Society did not accept a Spain reduced to a "mere" nation-state in its homeland. Therefore, its journal, from 1876 to 1914, dealt not only with Spain itself, but also with overseas colonies that Spain, at the time, had lost (e.g. Bolivia, Colombia), was losing (e.g. Cuba, Philippines), or reasserting control over (e.g. Guinea, Morocco). And the Geographical Society of Halle? Its main spatial format seems to have been neither a German "empire" nor "nation-state", but "regions" such as Thuringia and Saxony.[8]

[7]For around 1900, on colonialism in French schools: Manceron (2003); on the strong support from German schoolteachers for imperialism and nationalism: Wehler (1995): 1209.

[8]To be sure, if we gather certain structures, interests and activities of a Geographical Society from its journal or

7.5 Conclusion

The standardized method of recording journals of Geographical Societies from around the world that our research project has developed enables us, first, to profile their structures, activities, interests and evolution; second, to compare those profiles, and to observe connections among the Societies, in the form of exchanges of publications, invitations of other Societies' members, etc. Thus, we refute, for the long nineteenth century, the premise of previous research that Geographical Societies existed and operated in independence and isolation from one another. Especially, we overcome that research's methodological nationalism.

Even "provincial" Geographical Societies published considerable periodicals, where they did not just reprint materials from leading Societies such as Paris and London. By analyzing the journals, we shed light also on those smaller Societies, on which there is little to no published research. Equally, informative and surviving archives may be found for larger Societies rather than for smaller ones. We are nonetheless aware that the journal of a Society may not suffice to fully or correctly comprehend that Society; we would furthermore need its archives and research based on them (although the size of our sample prevents us from conducting such research). For example, the work of the Amsterdam Geographical Society became, from the 1890s onwards, more scientific, academic or professional—also to better support the Netherlands' colonialism in the Dutch East Indies, which at the time was becoming more active. In the Society's journal, though, we do not see more than a modified structure and, in its articles, a stronger focus on the Netherlands or its colonies (Schober 2018: 2, 8–9, 26–27, 30, 53–54, 56).

What the journals do not tell either—and perhaps not even archives would—is how the journals, and through them the issuing Societies, were received by the public. Who, if anyone, read those hundreds or thousands of pages per year? What impact did they have? Did the volumes, after all, do no more than decorate the libraries of other Geographical Societies? To find that out, we would need, for example, to record what other journals a journal cites in its own articles; and we would need various sources other than the journals.

From the rather qualitative data in our Word files, and the rather quantitative data in our Excel files, we may deduce a typology of Geographical Societies that so far has been recognized only in its rudiments. Among other things, the different types depend on the "spatial format(s)" into which a Society "spatializes" its world. Does it imagine and campaign for, or against, "nation-state" or "empire"; a combination of both, or something else? By "formatting" space in that way, the Societies, on the one hand, reacted to the unprecedented globalization during the "long" nineteenth century. On the other, they thereby acted on that globalization: as the Geographical Societies, from their cities, countries, empires and transnational networks around the world, supported certain spatial formats and opposed others, they co-determined, along with various other actors, which spatial formats rose to impose on at least parts of the globe a certain "spatial *order*", and which ones did not.

The Geographical Societies we have examined took up the challenge of conceiving the world as a whole. Our findings confirm that they constituted important fora for intellectual and social; political, military and economic elites to negotiate modes of respatialization that would be suitable for times of globalization. In the twentieth century, by contrast, the Societies lost their importance owing to their own success: thanks to their efforts to professionalize geography, the latter was established at universities and then at schools, where university graduates went as teachers. From then on, the "formatting" of the

further sources, this does not mean that all its members continuously supported those structures, shared those interests, or participated in those activities. Any Society's members may have been so heterogeneous that in practice, its structures, interests and activities may have been less uniform and consistent than what we read in the sources. However, if we are to look at many Societies at the same time, we cannot do without some generalizations.

globalized space, in national and other ways, took place at academic institutes and in classrooms, and Geographical Societies had to find other purposes.

References

Bader FJW (1978) Die Gesellschaft für Erdkunde zu Berlin und die koloniale Erschließung Afrikas in der zweiten Hälfte des 19. Jahrhunderts bis zur Gründung der ersten deutschen Kolonien. Erde 109:36–48

Barke M et al (2013) The Tyneside Geographical Society, 1887–1944: its rise and decline. Tyne & Tweed: J Assoc Northumberland Local History Soc 66:19–25

Bassin M (1983) The Russian Geographical Society, the "Amur Epoch", and the Great Siberian Expedition 1855–1863. Ann Assoc Am Geogr 73(2):240–256

Bell M et al (eds) (1995) Geography and imperialism, 1820–1940. Manchester University Press, Manchester

Bell M, McEwan C (1996) The admission of women fellows to the Royal Geographical Society, 1892–1914: the controversy and the outcome. Geogr J 162(3):295–312

Berelowitch W (1990) Aux origines de l'ethnographie russe. La Société de géographie dans les années 1840–1850. Cahiers du Monde russe et soviétique 31 (2–3):265–273

Bergevin J (1994) La société de géographie de Québec et la colonisation. In: Bruneau and Dory, pp 67–75

Bonnefont J-C (1999) La Société de géographie de l'Est. Une géographie militante et conviviale à la Belle Époque. Revue Géographique de l'Est 39(1). http://journals.openedition.org/rge/4551

Bruneau M, Dory D (eds) (1994) Géographies des colonisations, XVe–XXe siècles. L'Harmattan, Paris

Butlin RA (2009) Geographies of empire: European empires and colonies, c. 1880–1960. Cambridge University Press, Cambridge

Camboulives R (1981/82) Historique de la Société. Bulletin de la Société de Géographie de Toulouse 262—Supplement: Bulletin spécial de la Société de Géographie de Toulouse: Centenaire de la Société (1882–1982):7–13

Camu P (1957) Le quatre-vingtième anniversaire de la Société de géographie de Québec. Cahiers de géographie du Québec 2(3):135–140

Capel H (1981) Filosofía y ciencia en la geografía contemporánea. Barcanova, Barcelona

Carazzi M (1972) La Società Geografica Italiana e l'esplorazione coloniale in Africa (1867–1900). La Nuova Italia, Florence

Cerny T (2017) Für die nationale Sache mit Erdkunde? Langzeituntersuchung des Fachorgans der Tschechischen Geographischen Gesellschaft (1895–1920). Master's Thesis, Universität Leipzig

Clout H (2008a) Popular geographies and scholarly geographies in provincial France: The Société normande de géographie, 1879–1937. J Histor Geogr 34 (1):24–47

Clout H (2008b) Popular geographies in a French port city: the experience of the Le Havre Society of Commercial Geography, 1884–1948. Scott Geogr J 124(1):53–77

Driver F (2001) Geography militant: cultures of exploration and empire. Blackwell, Oxford

Ferrié J-N, Boëstch G (1996) La Société de géographie du Caire à l'époque coloniale. Mise en scène de la science et réseaux scientifiques. In: Petitjean P (ed) Les sciences coloniales. Figures et institutions. Orstom, Paris, pp 69–78

Friederichsen L (1898) Rückblick auf die Gründung und Entwickelung der Geographischen Gesellschaft in Hamburg während der ersten fünfundzwanzig Jahre ihres Bestehens. Mitteilungen der Geographischen Gesellschaft in Hamburg 14:1–43

Garcia JLL (2004) A Sociedade de Geografia de Lisboa e a propaganda colonial em Portugal no final do século XIX. Escola Superior de Educação da Guarda, Guarda

Godlewska A, Smith N (eds) (1994) Geography and empire. Blackwell, Oxford

Gómez Mendoza J (2005) Ciencia y movimiento colonial en la Sociedad Geográfica de París. Boletín de la Sociedad Geográfica Española 20:60–69

Greiner P (2017) Geographies of empire and nation in Scandinavia: Nordic Geographical Societies and their role in processes of spatialization and territorialization, 1876–1914. Master's Thesis, Roskilde University

Hayes E (2018) Geographical light: the Magic Lantern, the reform of the Royal Geographical Society and the professionalization of Geography, c. 1885–1894. J Histor Geogr 62:24–36

Heffernan M (2003) Histories of geography. In: Holloway SL et al (eds) Key concepts in geography. SAGE, London, pp 3–22

Hernández Sandoica E (1994) Colonialisme et sociétés géographiques dans l'Espagne du XIXe siècle. In: Bruneau and Dory, pp 23–34

Hooson D (ed) (1994) Geography and national identity. Blackwell, Oxford

Hudson B (1972) The New Geography and the New Imperialism: 1870–1918. Antipode 9(2):12–19

Jud P (1989) 100 Jahre Geographisch-Ethnographische Gesellschaft Zürich. Geographica Helvetica 44(3): 113–151

Klein J-F (2008) La Société de géographie de Lyon. Pour la Croix et la Soie? (1873–1908). In: Singaravelou, pp 94–111

Kollm G (1909) Geographische Gesellschaften, Zeitschriften, Kongresse und Ausstellungen. Geographisches Jahrbuch 32:409–438

Koner W (1878) Zur Erinnerung an das fünfzigjährige Bestehen der Gesellschaft für Erdkunde zu Berlin. Zeitschrift der Gesellschaft für Erdkunde zu Berlin 13:169–250

Kretschmer I, Fasching G (eds) (2006) Österreich in der Welt—die Welt in Österreich. Chronik der Österreichischen Geographischen Gesellschaft—150 Jahre (1856–2006). Österreichische Geographische Gesellschaft, Vienna

Kuitenbrouwer JJV (2004) The crowning chapter of geography: the Royal Scottish Geographical Society and the Scramble for Africa, 1884–1893. Itinerario 28 (2):135–156

Lejeune D (1993) Les Sociétés de géographie en France et l'expansion coloniale au XIXe siècle. Albin Michel, Paris

López-Ocón L (1994) Les activités scientifiques de la Société géographique de Lima. In: Bruneau and Dory, pp 95–106

López-Ocón L (1996) Les Sociétés de géographie: un instrument de diffusion scientifique en Amérique latine au début du XXe siècle (1900–1914). In: Petitjean P (ed) Les sciences coloniales. Figures et institutions, Orstom, Paris, pp 79–86

MacKenzie JM (1995) The provincial geographical societies in Britain, 1884–1914. In: Bell et al. pp 93–124

Manceron G (2003) École, pédagogie et colonies. In: Blanchard P, Lemaire S (eds) Culture coloniale. La France conquise par son empire, 1871–1931. Autrement, Paris, pp 92–103

Markham CR (1881) The fifty years' work of the Royal Geographical Society. John Murray, London

Marosi S (1994) A Magyar Földrajzi Társaság első 120 éve (1872–1992). Földrajzi Közlemények 118/42 (1):5–34

Meyer JW et al (1997) World society and the nation-state. Am J Sociol 103(1):144–181

Middell M (2019) Raumformate—Bausteine in Prozessen der Neuverräumlichung = Working Paper Series des SFB 1199 an der Universität Leipzig, no. 14. Leipziger Universitätsverlag, Leipzig

Milanini Kemény A (1973) La Società d'esplorazione commerciale in Africa e la politica coloniale (1879–1914). La Nuova Italia, Florence

Monina G (2002) Il consenso coloniale. Le Società geografiche e l'Istituto coloniale italiano (1896–1914). Carocci, Rome

Natili D (2008) Un programma coloniale: la Società Geografica Italiana e le origini dell'espansione in Etiopia (1867–1884). Gangemi, Rome

Nicolaï H (1994) Les géographes belges et le Congo. In: Bruneau and Dory, pp 51–67

Nogué J, Villanova JL (2002) Spanish colonialism in Morocco and the Sociedad Geográfica de Madrid, 1876–1956. J Histor Geogr 28(1):1–20

Payne MM (1963) 75 years exploring earth, sea and sky: National Geographic Society observes its diamond anniversary. National Geographic Magazine 123(1): 1–43

Péaud L (2016) La géographie, émergence d'un champ scientifique. France, Prusse et Grande-Bretagne (1780–1860). ENS Éditions, Lyon

Péhaut Y (1994) Géographie, colonies et commerce à Bordeaux, 1874–1939. In: Bruneau and Dory, pp 77–94

Rodríguez Esteban JA (1996) Geografía y colonialismo. La Sociedad Geográfica de Madrid (1876–1936). Universidad Autónoma de Madrid, Madrid

Schober V (2018) Die Koninklijk Nederlands Aardrijkskundig Genootschap im Zeitalter des Imperialismus, 1873–1914. Bachelor's Thesis, Universität Leipzig

Schou A (1977) Det Kongelige Danske Geografiske Selskab, 1876–1976 (Foredraget på festmødet den 7. december 1976 i anledning af 100 års jubilæet). Geografisk Tidsskrift 76:X–XIV

Schulten S (2001) The geographical imagination in America, 1880–1950. University of Chicago Press, Chicago

Singaravelou P (ed) (2008) L'empire des géographes. Géographie, exploration et colonisation, XIXe–XXe siècles. Belin, Paris

Société royale belge de géographie (ed) (1926) Cinquantième anniversaire de la fondation de la Société royale belge de géographie. Bulletin de la Société royale belge de géographie 50(3–4):173–194

Steinbach-Hüther N et al. (2019) Geographiegeschichtsschreibung und Digital Humanities. Neue Methoden für Zeitschriftenanalysen = Working Paper Series des SFB 1199 an der Universität Leipzig, no. 15. Leipziger Universitätsverlag, Leipzig

Stintzing M (2017) Geographische Gesellschaften im 19. Jahrhundert. Ein Vergleich der Aktivitätsprofile der (Real) Sociedad Geográfica de Madrid und der Sociedad Geográfica de Lima. Master's Thesis, Otto-Friedrich-Universität Bamberg

van der Velde P (1995) The Royal Dutch Geographical Society and the Dutch East Indies, 1873–1914: from colonial lobby to colonial hobby. In: Bell et al. pp 80–92

Villanova Valero JL (1999) La Sociedad Geográfica de Madrid y el colonialismo español en Marruecos (1876–1956). Documents d'Anàlisi Geográfica 34:161–187

Wehler H-U (1995) Deutsche Gesellschaftsgeschichte, vol. 3: 1849–1914. C.H. Beck, Munich

Weiss C (2007) Wie Sibirien „unser" wurde. Die Russische Geographische Gesellschaft und ihr Einfluss auf die Bilder und Vorstellungen von Sibirien im 19. Jahrhundert. V&R unipress, Göttingen

Zusman P (1996) Sociedades geográficas na promoção do saber ao respeito do território: estratégias políticas e acadêmicas das instituições geográficas na Argentina (1879–1942) e no Brasil (1838–1945). Dissertation, Universidade de São Paulo

Personified Continents in Public Places: Internationalism, Art, and Geography in Late Nineteenth Century Paris

8

Toshiyuki Shimazu

Abstract

This paper examines the yet unexplored relationships between internationalism, art, and geography by discussing the two cases of Paris, in the late nineteenth century. Using a landscape-as-text approach and the concepts of scale jumping and imaginative geographies, I focused here on the female statues representing the continents. Late nineteenth century Paris was characterized as the site where occurred a host of projects and events promoting internationalism by means of some kind of scale jumping. The world's fairs and international congresses were typical events concerning internationalism, especially a hegemonic one. The erection of *Les Quatre Parties du Monde* was virtually linked to the debates and assertions surrounding the selection of the prime meridian, which was one of the themes at the two Paris international geographical congresses: the Congrès International des Sciences Géographiques in 1875, and the Congrès International de Géographie Commerciale in 1878. The physiognomies of *Les Quatre Parties du Monde* delivered an apparent message of the supremacy of Europe over others, and the fountain and statues were intended to symbolize the Paris meridian itself and to assert its status as the starting point for the calculation of the world's longitude. The group of *Les Six Continents* was stemmed from the sort of hegemonic internationalism and imaginative geographies that found themselves at the 1878 Paris International Universal Exposition. While the appearances of *Les Six Continents* indicated a linear progress from primitive Oceania to civilized Europe, their arrangement at the inaugural ceremony at the Palais du Trocadéro, signified the historical and geographical centrality of the Mediterranean world. Both *Les Quatre Parties du Monde* and *Les Six Continents* could be taken as the descendants of the title page of Ortelius' first world atlas. The practice to erect the statues of the continents could be regarded as a geographical practice, and these statues could also be understood as a different kind of geo-body.

Résumé

Cet article examine les relations encore inexplorées entre l'internationalisme, l'art et la géographie en discutant les deux cas de Paris à la fin du XIXe siècle. En utilisant une approche du paysage en tant que texte et les concepts de saut d'échelle et de géographies imaginatives, je me suis concentré ici sur les statues féminines représentant les continents. A la fin du XIXe siècle, Paris a été caractérisée

T. Shimazu (✉)
Wakayama University, Wakayama, Japan
e-mail: shimazu@wakayama-u.ac.jp

B. Schelhaas et al. (eds.), *Decolonising and Internationalising Geography*, Historical Geography and Geosciences, https://doi.org/10.1007/978-3-030-49516-9_8

comme le site où se sont produits une multitude de projets et d'événements promouvant l'internationalisme à travers une sorte de saut d'échelle. Les expositions internationales et les congrès internationaux sont des événements typiques concernant l'internationalisme, en particulier celui hégémonique. L'érection des *Quatre Parties du Monde* était virtuellement liée aux débats et affirmations autour de la sélection du premier méridien, qui fut l'un des thèmes des deux congrès géographiques internationaux de Paris: le Congrès International des Sciences Géographiques en 1875 et le Congrès International de Géographie Commerciale en 1878. Les physionomies des *Quatre Parties du Monde* ont véhiculé un message apparent de la suprématie de l'Europe sur les autres, tandis que la fontaine et les statues étaient destinées à symboliser le méridien de Paris lui-même et à affirmer son statut de point de départ pour le calcul de la longitude du monde. Le groupe des *Six Continents* est issu du type d'internationalisme hégémonique et de géographies imaginative qui se sont retrouvés à l'Exposition universelle internationale de Paris de 1878. Alors que les figures des *Six Continents* indiquent un progrès linéaire de l'Océanie primitive à l'Europe civilisée, leur arrangement lors de la cérémonie inaugurale au Palais du Trocadéro signifiait la centralité historique et géographique du monde méditerranéen. *Les Quatre Parties du Monde* et *Les Six Continents* pourraient tous deux être considérés comme les descendants de la page de titre du premier atlas mondial d'Ortelius. La pratique d'ériger les statues des continents pourrait être considérée comme une pratique géographique et ces statues pourraient également être comprises comme un type différent de géo-corps.

Resumen

Este artículo examina las relaciones aún inexploradas entre internacionalismo, arte y geografía, discutiendo los dos casos de París al final del siglo XIX. Utilizando un abordaje de paisaje como texto y los conceptos de salto de escala y de geografías imaginativas, me concentré aquí en las estatuas femeninas que representan los continentes. A fines del siglo XIX, París se caracterizó como el lugar donde se produjeron una serie de proyectos y eventos que promovieron el internacionalismo mediante algún tipo de salto de escala. Las exposiciones universales y los congresos internacionales fueron eventos típicos relacionados con el internacionalismo, especialmente lo hegemónico. La construcción de *Les Quatre Parties du Monde* estuvo virtualmente vinculada a los debates y afirmaciones que rodearon la selección del primer meridiano, que fue uno de los temas en los dos congresos geográficos internacionales de París: el *Congrès International des Sciences Géographiques* en 1875 y el *Congrès International de Géographie Commerciale* en 1878. Las fisionomías de *Les Quatre Parties du Monde* transmitieron un mensaje aparente de la supremacía de la Europa sobre los demás, mientras la fuente y las estatuas tenían la intención de simbolizar el meridiano de París y afirmar su estatus como punto de partida para el cálculo de la longitud del mundo. El grupo de *Les Six Continents* surgió del tipo de internacionalismo hegemónico y geografías imaginativas que se encontraron en la Exposición Universal Internacional de París de 1878. Si bien las apariciones de *Les Six Continents* indicaron un progreso lineal desde la primitiva Oceanía hasta la Europa civilizada, su disposición en la ceremonia inaugural en el *Palais du Trocadéro* significó la centralidad histórica y geográfica del mundo mediterráneo. Tanto *Les Quatre Parties du Monde* como *Les Six Continents* podrían considerarse descendientes de la portada del primer atlas mundial de Ortelius. La práctica de erigir las estatuas de los continentes podría considerarse una práctica geográfica y estas estatuas también podrían entenderse como un tipo diferente de geo-cuerpo.

Keywords

History of geography · Landscape-as-text · Imaginative geographies · Scale jumping · Hegemonic internationalism · Allegorical statue · Prime meridian · World's fair

8.1 Introduction

Geography had long been seen as engaging in the task of describing and representing the earth as a whole since Ptolemy's *Geographia*. The interrelations between the subdivisions of the earth had also entered into the realm of geographers' interest. Since the late eighteenth century, this subdividing has been largely undertaken by modern nation-states and their colonies. Consequently, international relations through the flow of people, goods, money, information, and practices have been included in the topics in the branches of human geography, especially in political geography. In addition to this, the relationships between geography and internationalism as thought and practice are becoming an emerging topic in the most recent scholarship in political geography, historical geography, and the history of geography (Hodder et al. 2015; Shimazu 2015; Ferretti 2016). Internationalism as a conceptual rather than a historically used term could be defined as the thought and practice, which celebrates transnational consciousness and advocates pan-regional or global cooperation. Therefore, some thought that geography, in general, was instrumental to the teaching and dissemination of internationalism (Otlet 1911; Ferretti 2016). Moreover, it should be noted, that "internationalism has both a history and a geography" (Hodder et al. 2015: 3). In particular, the fact deserves attention that internationalism has been bound to specific time and space, and also embodied in various material forms. Hodder et al. (2015: 5) proposes seven future research directions for the historical geographies of internationalism, the last of which is concerned with the question "how the international was articulated visually, and specifically through cartography and related forms of visual culture." This is quite suitable for my present research interest. Among visual culture, my focus here is on art, especially sculptures and statues, which are themselves part of the landscape. They are, therefore, also part of what geographers have long tried to describe, explain, and interpret.

This paper examines along these lines the yet unexplored relationships between internationalism, art, and geography. Grataloup (2009: 116) pointed out that the symbolism of women as parts of the world had often been utilized in statue forms through nineteenth century Europe. Here special attention is paid to the installation of allegorical female statues representing the continents of the world in late nineteenth century Paris. Two cases will be discussed in the following. One is *Les Quatre Parties du Monde*, located on the Avenue de l'Observatoire at the southern end of the present Jardin des Grands Explorateurs Marco-Polo et Cavelier-de-la-Salle (Fig. 8.1). The other is a group of six separate statues symbolizing the six continents, which have been installed on the parvis, in front of the Musée d'Orsay, since its opening in 1986, and often been called *Les Six Continents* (Dupre 1991: 8; Pingeot 1994: 5). These statues were, however, initially installed at the Palais du Trocadéro, which was one of the main venues for the 1878 Paris International Universal Exposition (Figs. 8.3 and 8.4).

The connection between internationalism and art has already been discussed in art history and art studies, many of which put stress on nationalist or Eurocentric power relations inherent in the assertion of internationalism in the art scene (Judd 1992; Araeen 1994; Jones and Clark 2008; Brockington 2009). Halliday (1988) once distinguished three types of internationalism: liberal, hegemonic, and radical internationalism. According to this, internationalism in the nineteenth century could be characterized in most cases as a hegemonic one with a few exceptions. Internationalism and related events discussed in the following will be also of the same kind, and these include the international geographical congresses held in Paris.

The methodology I will employ here is a "landscape-as-text" approach, which regards landscapes or material objects in space as

Fig. 8.1 Les Quatre Parties du Monde (1874). *Source* L'Univers Illustré, 17e Année, N° 1020, 10 Octobre 1874, p. 648

conveying meaning like oral and literary texts (Bellentani 2016). This approach was once fully discussed by Duncan (1990), which considered the landscape as "a signifying system" and this semiotic point of view still deserves attention. The landscape is not a passive reflection of past culture, but an active signifier conveying meanings and thereby playing various roles in social and political processes. As part of the landscape, plastic arts have been deployed as such a signifier through which various meanings and attitudes could easily be transmitted to the wider public. As for hegemonic internationalism, it is worth noting that the above-mentioned Paris statues, could have transmitted to the public distinctive "imaginative geographies." The concept of imaginative geographies originated from the renowned literary critic Edward Said (1935–2003) and means "representations of peoples and places that express the perceptions, desires, fantasies, fears, and projections of their authors, who are generally external observers" (Desbiens 2017: 1). Asymmetric power relations have always been contained in imaginative geographies. Then, what sort of statues were erected in Paris, as a consequence of the interplay between internationalism, art, and geography? What sort of imaginative geographies were conveyed through those statues? These are my key research questions in the following sections.

8.2 Internationalism in Late Nineteenth Century Paris

Before proceeding to case studies, I briefly touch upon the burgeoning of internationalism in late nineteenth century Paris. No need to say, Paris experienced a massive urban renewal under the leadership of Baron Georges-Eugène Haussmann (1809–1891), prefect of the Seine under the Second Empire appointed by Emperor Napoleon III (1808–1873). During the times of Haussmann and after, Paris was transformed to be what is later referred to as the "capital of the nineteenth century" (Benjamin 1969) or the "capital of modernity" (Harvey 2003).

This urban renewal and transformation reflected the Second Empire's ambition to make Paris a privileged site for a version of "scale jumping," that is, "turning local into regional, national, and global movements, escaping the traps of localism, parochialism, and particularism through an expansion of geographic and political reach" (Jones et al. 2017: 143). The real or imaginary extension of the spatial reach of political, economic, social, and cultural activities has usually been conducted to gain more and more profit, power, authority, and legitimacy. For Paris, like other cities such as London, or Berlin, this scale jumping meant an international shift in the scale on which a variety of its activities unfolded. With regard to the female statues mentioned above, one example of an imaginary scale jumping is found in the huge main façade of the Gare du Nord, which was first inaugurated in 1846. The present station building, designed by the French architect Jacques Ignace Hittorff (1792–1867), was constructed between 1861 and 1865 (Rice 2016: 27) and its southern façade was furnished with twenty-three female statues representing European cities including Paris itself. The statue of Paris, seems most elegant, was at the center of the rooftop. Other eight statues, standing on both sides of that of Paris, in order from east to west end, represented Cologne, Berlin, Vienna, London, Brussels, Warsaw, Amsterdam, and Frankfurt, respectively. The Gare du Nord served as the main gateway to these eight cities. Residual fourteen statues representing French cities were set in the middle range of the façade. This kind of arrangement obviously reflected an intention to assert Paris' superiority and reign over other European cities (Keilo 2015; Fratello 2017), and moreover, reflected imaginative geographies from Paris. Those female statues were materially mobilized in this imaginary scale jumping.

As Harvey (2003: 114–115) put it correctly, this "new world of spatial interconnections facilitated by modern networks of communication and materialized through commodity exchange" was destined to become "the phantasmagoria of universal capitalist culture." Nevertheless, from the point of view of internationalism in general, notable events took place in late nineteenth century Paris. Among others, Paris was the most privileged site for the world's fair. It took place five times in late nineteenth century Paris (in 1855, 1867, 1878, 1889, and 1900). Apparently, the world's fair was one of the earlier manifestations of internationalism on the global scale and, as later discussed in relation to the 1878 International Universal Exposition, the material landscapes of the world's fair could themselves be understood as the vehicle of imaginative geographies deriving from the imagining subject.

Another symptom of the rise of global internationalism in the late nineteenth century Paris, was the holding of international congresses, especially scientific ones. From a geographical point of view, it is worth noting that "science more generally was increasingly internationalized during the late nineteenth century" (Withers 2017: 139). This meant that another version of scale jumping was needed and conducted for science and scientific knowledge to achieve a more universalized or verified status. On the other hand, scientific internationalism also tended to flourish in specific regions or sites and was prone to be a hegemonic one in most cases. Fuchs (2002: 205) stressed that the "internationalization of science and scientific institutions is a product of nineteenth century Europe." The number of international congresses increased from the 1850s onwards, and Paris turned out to be the most privileged site for the venues of international congresses and the headquarters of international organizations. Between 1840 and 1913, a total of 678 international congresses were held in Paris, followed by 272 for Brussels, and 242 for London (Lyons 1963: 16–17; Tapia and Taieb 1976: 17). What must not be forgotten here is that there was a strong connection between international congresses and the Paris world's fairs. Between 1855 and 1900, five world's fairs held in Paris, attracted a total of 369 international congresses to the same city, and this explained 84.8 percent of the total 435 international congresses held in the corresponding five years (Tapia and Taieb 1976: 15). As Lyons (1963) observed, Paris had long been the cultural and intellectual center of Europe, and the French language still held the status of international language in those times. In addition to this, an important fact is that this tendency was strengthened by the French government itself, which "deliberately offered exceptional facilities to the organizers of congresses" (Lyons 1963: 17). International congresses were thus understood as the "scientific and intellectual dimensions of the material enterprise" (Rasmussen 1989: 24). The French government managed to maintain its capital city's international superiority by assisting the organizers of various scale-jumping projects who themselves intended to pursue their own goals and interests.

8.2.1 Les Quatre Parties Du Monde (1874)

The first case I want to discuss is *Les Quatre Parties du Monde*, which comprises four bronze female statues representing Europe, America, Africa, and Asia, respectively. They were sculptured by Jean-Baptiste Carpeaux (1827–1875), and were all in the form of an almost naked woman. Their realistic appearances thus invited some critical voices on the occasion of their first exhibition at the Salon of 1872, at the Palais de l'Industrie (Margerie 2014: 166). The statue of Europe is characterized by her wind-blown hair. That of America wears a feather headband reminiscent of the Native Americans. That of Africa has a distinctive afro hairstyle and two large earrings. Lastly, that of Asia looks like a Chinese woman with a Manchu pigtail hairstyle, which had been, in fact, a male style and was probably misunderstood as a female one. While America, Africa, and Asia have accessories or distinctive hairstyles, Europe has no such elements and assumes a touch of divineness. The contrast between unmarked Europe and marked others are evident, and discernible from there is a type of imaginative geographies from the West.

These statues are together lifting an armillary sphere overhead. A smaller terrestrial globe is set inside the armillary sphere. The whole statue is installed on a circular stone pedestal, which is surrounded by a large fountain. The entire structure was completed in August 1874 (Margerie 2014: 157), and *L'Univers Illustré*, one of the Paris illustrated newspapers, reported about the fountain with an illustration on October 10th, 1874, two months after its inauguration (Fig. 8.1). While the fountain was initially named *Fontaine de l'Avenue de l'Observatoire* (Alphand 1867–1873: 235; Darlet 1874: 648), it has later been called in various ways, such as *Fontaine Carpeaux*, *Fontaine des Quatre-Parties-du-Monde,* and so on (Bibliothèque de l'INHA 2002).

Under the urban renewal policy of Napoleon III and Baron Haussmann favoring large straight boulevards, extensive gardens and decorated fountains, the fountain and statues were originally conceived in 1867, by Gabriel Davioud (1824–1881), architect-in-chief for the Service des Promenades et Plantations. This conception marked one of the final stages in the redevelopment and landscaping projects for the Avenue de l'Observatoire, running south and north from the Observatoire de Paris to the Jardin du Luxembourg. Originally, this avenue had been laid out exactly along the Paris meridian, and had been recognized as the materialization of that imaginary line. Davioud commissioned Carpeaux to produce the statuary to be installed on the fountain, which was itself intended to symbolize the site's imaginary geographical centrality. Carpeaux was himself a favorite with Napoleon III and had already produced *La France éclairant le Monde et protégeant la Science et l'Agriculture*, magnificent stone statuary installed on the pediment above the southern façade of the refurbished Pavillon de Flore (Muller 1867). This included a female statue with a torch representing Imperial France, and a male statue with a terrestrial globe and a compass representing science (Anonymous 1875). The latter male statue might be regarded as a geographer in that a person with a globe and a compass was sometimes referred to as an allegory of geography (Sterling and Francine Clark Art Institute 2011). And it should be added that the Pavillon de Flore became one of the venues for the "Exposition de géographie" held in conjunction with the Congrès International des Sciences Géographiques in 1875 (see below) (Société de Géographie 1878: iii–v).

Although *Les Quatre Parties du Monde* was sculptured largely in the same line as *La France éclairant le Monde et protégeant la Science et l'Agriculture*, an important difference is that the former was specifically intended to symbolize the Paris meridian itself as a ruling center over the whole world. In relation to this, it could be said that the entire redevelopment scheme for the Avenue de l'Observatoire was closely linked to the controversy and competition over the status of the prime meridian in those times. Therefore, it would be helpful to situate the installation of *Les Quatre Parties du Monde* within the context of the international conflict over the position of "zero degrees" (Withers 2017).

In the late nineteenth century, the determination of a single prime meridian became a major issue in the emerging international scientific communities, including especially that of geography. Eugène Cortambert (1805–1881), French geographer and one of the principal members of the Société de Géographie de Paris, lamented that "each nation wants to have this initial meridian, passing through its capital city or its principal observatory" (Cortambert 1879a: 199). As a consequence, multiple prime meridians coexisted simultaneously, and they were competing with each other (Fig. 8.2). Especially, one of the harshest competitions took place between Paris and Greenwich. One of the arenas for debate was the International Geographical Congress, which was first held in Antwerp, in 1871. The Antwerp congress, which was named "Congrès International pour le Progrès des Sciences Géographiques, Cosmographiques et Commerciales," listed 600 names as "adherents" (Anonymous 1872a). The number of actual participants is estimated at about 300 (Kish 1989). At the congress, scientific discussions were classified into four major sections: (I) Geography, (II) Cosmography, (III) Navigation, Voyages, Commerce, Meteorology, Statistics, and (IV) Ethnography. In each section, discussions were made on the questions assigned in advance. While the question "Could we not agree to adopt the same first meridian?" was originally assigned to the Cosmography section, it was discussed also in the Geography section and in the general sessions. The majority of congress participants supported the meridian of Greenwich as the single prime meridian, and in a general session the following resolution was adopted: "The Congress expresses the wish that, for maritime route maps, be adopted a common first meridian, which would be that of the Observatory of Greenwich…As for terrestrial and coastal maps, the Congress thinks that there are more advantages than disadvantages for each State to maintain its particular meridian." In relation to this resolution, Émile Levasseur, one of the founders of the Société de Géographie Commerciale de Paris (1873), stated immediately, that "if we were in the seventeenth or eighteenth century, it would be appropriate to adopt the meridian of Paris. But the circumstances have changed…The most used nautical charts, especially for distant expeditions, are of English origin…So let no one be mistaken about the meaning of this decision; it does not establish at all that the chosen meridian is better in itself… For the terrestrial maps, we found it good to maintain the status quo" (Anonymous 1872b: 254–256). While Levasseur admitted the advantage of the Greenwich meridian for nautical charts, he also maintained the significance of the Paris meridian for the terrestrial maps made by the French.

In August 1875, just one year after the installation of *Les Quatre Parties du Monde*, the second International Geographical Congress, named "Congrès International des Sciences Géographiques" was held in Paris, under the sponsorship of the Société de Géographie de

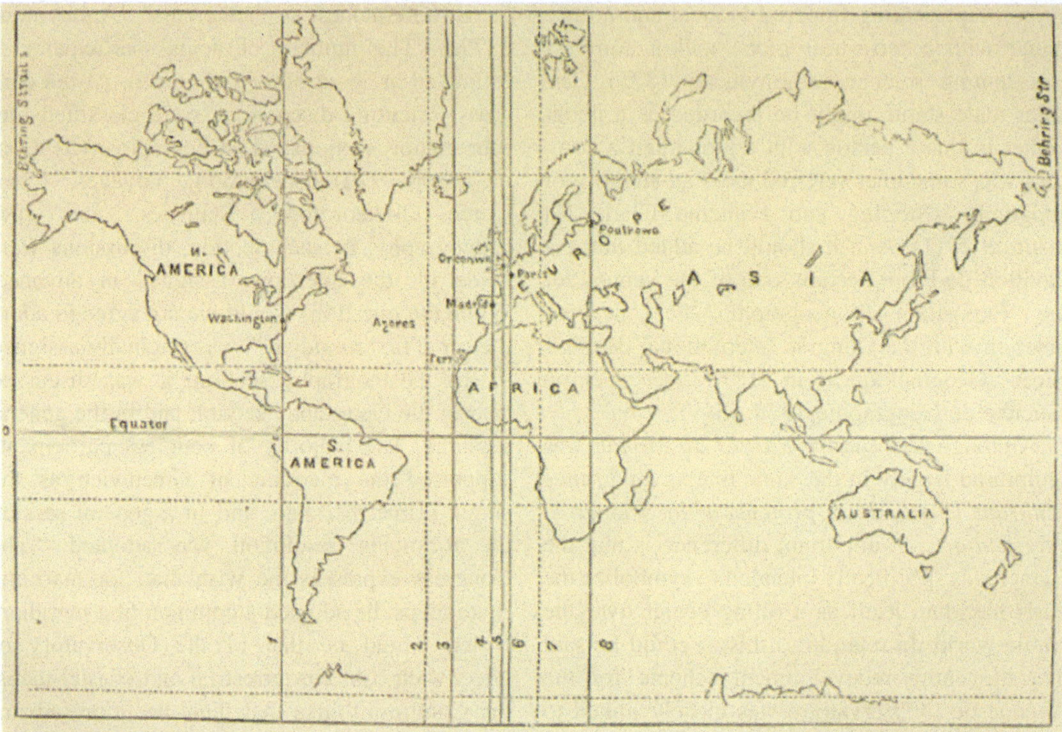

Fig. 8.2 First Meridians—1 Washington; 2 St. Michael; 3 Ferro; 4 Madrid; 5 Greenwich; 6 Paris; 7 Bouthillier de Beaumont's; 8 Poulkowa. *Source* Cortambert (1879b: 157)

Paris. The number of participants was reported as 1477 (Kish 1989). The committee of the congress was divided into five sections: I. Scientific Section, II. Section of Organization, III. Section of Exposition, IV. Section of Publicity, and V. Section of Accounting. Moreover, the Scientific Section was further divided into seven groups: Group I (mathematical geography, geodesy, topography), Group II (hydrography, maritime geography), Group III (physical geography, general meteorology, general geology, botanical and zoological geography, general anthropology), Group IV (historical geography, history of geography, ethnography, philology), Group V (economic, commercial, and statistical geography), Group VI (education and diffusion of geography), and Group VII (explorations, scientific, commercial, and sightseeing voyages). Like the Antwerp congress, each group was given related questions in advance. While the issue of the prime meridian was not initially included in these preset questions, it was indeed discussed mainly in Group I and II. In particular, at a discussion in Group I, Aimé Laussedat, member of the Société de Géographie de Paris, proposed that the Island of Ferro as 20° of longitude west of the Paris meridian was favorable for a prime meridian. Most members in Group I voted for this proposal (Société de Géographie 1878: 30). Consequently, one of the resolutions from Group I was as follows: "the conventional meridian of the Island of Ferro adopted by the old geographers should be recommended to all publishers of geographical maps, as the meridian-origin, from which will be counted the longitudes" (Société de Géographie 1880: 400). In fact, the Ferro meridian had been declared as the prime meridian by the King of France Louis XIII (1601–1643), in 1634, under the initiative of Cardinal Richelieu (1585–1642) (Bruzen de la Martinière 1768: 221). Therefore, it can be said that this resolution was virtually supporting the superiority and functionality of the Paris meridian itself.

In 1879, Eugène Cortambert wrote a short article on the prime meridian issue in the French popular science magazine *La Nature*. He deplored that an "over-patriotic selection by the English and French was a bad example" (Cortambert 1879a: 199). However, after all, he supported "the meridian that is passing through the strait of Behring and through 10° east of Paris" (Cortambert 1879a: 200). This prime meridian had originally proposed by Henry Bouthillier de Beaumont (1819–1898), president of the Société de Géographie de Genève, at the Congrès International de Géographie Commerciale held in Paris, in 1878 (Bouthillier de Beaumont 1881: 212) (Fig. 8.2). This congress was one of the international congresses held in connection with the International Universal Exposition in the same year (see previous and later sections). Cortambert (1879a: 200) overtly stated that "the initial meridian just 10° east of Paris" would facilitate and prompt "the conversion from determinations established on Paris, and the Island of Ferro." His article was soon translated into the American magazine *Popular Science Monthly* of the same year (Cortambert 1879b). An element of patriotic propaganda cannot be concealed and it may be said that he did the same thing as what he himself criticized in the same article.

These international geographical congresses all could be characterized as the intellectual arena in which nationalism and internationalism were confronted with each other and the former was sometimes covertly promoted under the banner of the latter. It could be said that the construction of the *Fontaine de l'Avenue de l'Observatoire* and the installation of *Les Quatre Parties du Monde* were closely linked to the sort of hegemonic internationalism that found expression at those international geographical congresses.

8.2.2 Les Six Continents (1878)

Next, I want to discuss the second case, that is, a group of six female statues, which has often been called *Les Six Continents* and has received visitors at the parvis of the Musée d'Orsay since 1986. This comprises six statues representing, respectively, Europe, Asia, Africa, North America, South America, and Oceania. As previously mentioned, these female statues were initially made "in order to correspond with the universal character of the Exposition" (Exposition Universelle de 1878: 126). The Commission of Fine Arts for the 1878 Paris International Universal Exposition, appointed by the Minister of Agriculture and Commerce, entrusted "six of the most eminent artists of the modern French school" with the task (Exposition Universelle de 1878: 126). *L'Europe* was undertaken by Alexandre Schoenewerk (1820–1885). She puts a laurel wreath on her head, having a shield and books at both sides. The most important feature is that she is the only one fully dressed among six statues. *L'Asie* was undertaken by Alexandre Falguière (1831–1900). She wears a Japanese coiffure, having a small Hindu statue on her knee and an elephant behind. In contrast to *L'Europe*, her breast is half exposed. Similarly, half exposed is the breast of *L'Afrique*, which was sculptured by Eugène Delaplanche (1836–1891). She carries a round basket with vegetables or fruit. The statues of residual three new continents are further characterized against those of three old continents by their fully exposed breasts. *L'Amérique du Nord* was made by Ernest Eugène Hiolle (1834–1886). She puts a large necklace around her neck, with a huge axe in her hand. *L'Amérique du Sud* was sculptured by Aimé Millet (1819–1891). Like *L'Amérique du Nord*, she also wears a double necklace, with a shield in her hand and a pile of fruit at her feet. Lastly, *L'Oceanie* was completed by Mathurin Moreau (1822–1912). Putting a hairband around her head and accompanied by a kangaroo, she assumes a more primitive appearance reminiscent of an Aborigine woman. Since 1986, they have been aligned from *L'Oceanie* on the right end to *L'Europe* on the left end, simply indicating a linear progress from the primitive to the civilized (Dupre 1991). A type of Eurocentric imaginative geographies can be seen there. However, the way in which these female statues were arranged at

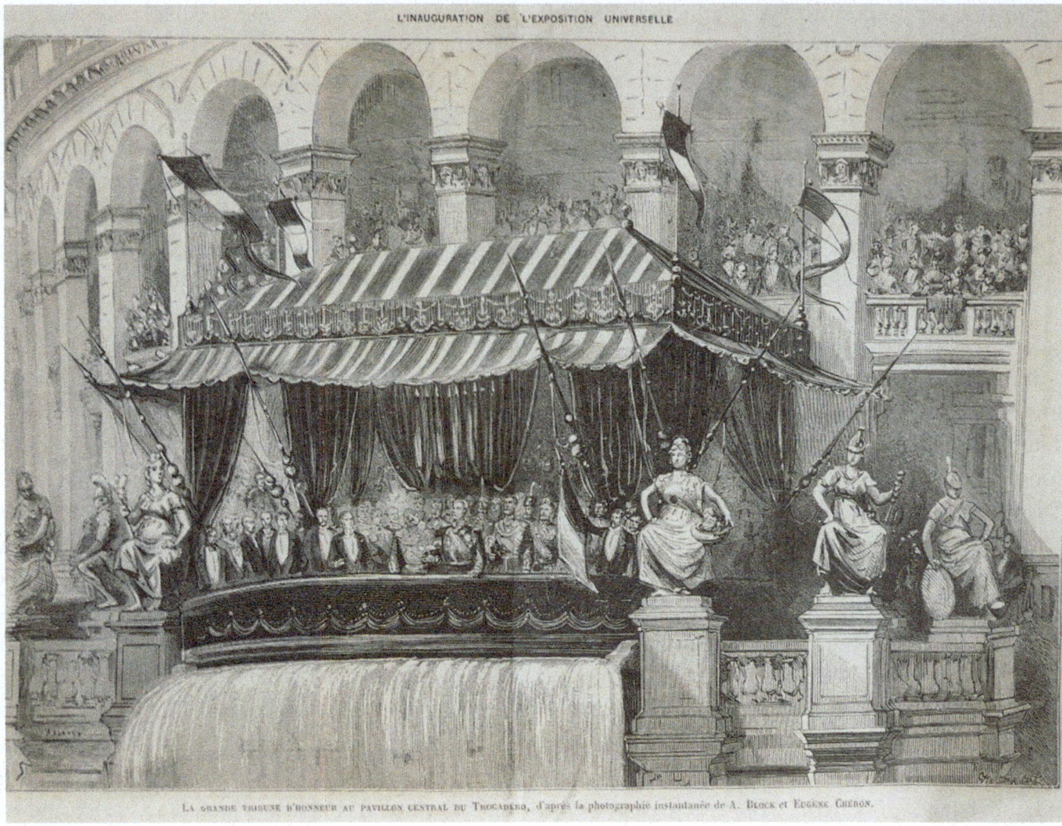

Fig. 8.3 Inauguration of the International Universal Exposition (1878). *Source* L'Illustration, 36ᵉ Année, Vol. LXXI, N° 1837, 11 Mai 1878, p. 300

the 1878 International Universal Exposition was fairly different from such a linear, simplistic one. At the inaugural ceremony at the Palais du Trocadéro, they were arranged circularly around the central stage for the ceremony (Fig. 8.3).

In order to explain this point further, it would be helpful to touch upon the form, function, and meaning of the Palais du Trocadéro itself (Fig. 8.4). This massive edifice was constructed originally as one of the major pavilions for the 1878 International Universal Exposition, and was finally demolished in order to construct the Palais de Chaillot for the 1937 International Exposition of Art and Technology in Modern Life. It was the same architect as in the case of *Fontaine de l'Avenue de l'Observatoire* who chiefly undertook the construction of the Palais du Trocadéro. Gabriel Davioud and his collaborator Jules Bourdais (1835–1915), won the competition and constructed their grand edifice on the hill of Chaillot. It became the only one pavilion, which survived after the 1878 Exposition as a museum and conference halls (Destors 1881: 11–12). A round shaped main building was on the center, with two long curved galleries stretching from both sides. Attached to the main building was an open terrace, which served as the central stage for the inaugural ceremony (Figs. 8.3 and 8.4). During the Exposition, this main building served as the venue for many international congresses including the Congrès International de Géographie Commerciale previously mentioned (Anonymous 1878).

It should be noted that the Palais du Trocadéro was itself conceived as the central pivot of the whole world. This interpretation could be evidenced by an extraordinarily impressive drawing that appeared on the cover page of

Fig. 8.4 Palais du Trocadéro (circa 1878). Backside description: "ALEXANDRE QUINET PHOTOGRAPHE DU CONGRES ET DE LA SOCIETE DE GEOGRAPHIE DE PARIS". *Source* Author's collection

L'Illustration on May 4th, 1878 (Fig. 8.5). It was drawn by Louis Morin (1855–1938), and entitled *Le 1er Mai 1878*, when the inaugural ceremony for the Exposition occurred. Ships, trains, peoples, and animals, all of these are pilgrimaging towards the Palais du Trocadéro, on an imaginary divine hilltop. It could be seen as a manifestation of a typically hegemonic internationalism and also of imaginative geographies from Paris, reminiscent of the group of female statues at the Gare du Nord (see previous section). And it could be also said that an imaginary scale jumping from the local to the global scales was conducted there.

It is in this context that the drawing of the inaugural ceremony could be fully understood (Fig. 8.3). It was drawn by Charles Baude (1853–1935), and appeared in *L'Illustration* on May 11th, 1878. The terrace attached to the round shaped main building is covered with a striped square canopy. There is an artificial waterfall under the central stage, which is surrounded by six female statues. *L'Europe* is placed on the near left side of the central stage and *L'Afrique* is on the near right side of the same stage. *L'Amérique du Nord* is on the right side of *L'Afrique*. And on the rear right side of *L'Amérique du Nord* is placed *L'Amérique du Sud*. On the opposite side, *L'Asie* is placed on the front left side of *L'Europe,* and on the rear left side is placed *L'Oceanie*. This arrangement of female statues would mean that the central stage was fabricated itself as a miniature whole world. The central stage was, as it were, the Mediterranean. The left side from *L'Europe* was the eastern world (*L'Asie* and *L'Oceanie*) and the right side from *L'Afrique* was the western world (*L'Amérique du Nord* and *L'Amérique du Sud*). These female statues were placed according to the geographical or psychological distances from the Mediterranean world. This interpretation could be supported by the neo-Byzantine style of

Fig. 8.5 Le 1er Mai 1878 (Dessin allégorique de M. Morin). *Source* L'Illustration, 36e Année, Vol. LXXI, N° 1836, 4 Mai 1878, p. 281

the Palais du Trocadéro, and the fact that the site was where Emperor Napoleon I had once projected for his son future Napoleon II a new palace complex named the "palais du roi de Rome" in the early nineteenth century (Exposition Universelle de 1878: 25).

Messages from *Les Six Continents* at the Palais du Trocadéro are, unlike those at the Musée d'Orsay, not straightforward ones (Fig. 8.3). Like Morin's drawing, hegemonic internationalism and Eurocentric imaginative geographies are certainly deciphered from there. However, the juxtaposition of *L'Europe* and *L'Afrique* on both sides of the central stage meant the celebration of the tradition and civilization of the Mediterranean world. While *L'Amérique du Sud* and *L'Oceanie* are pushed into the periphery, *L'Amérique du Nord* and *L'Asie* occupy the symmetrical position on both sides of the center stage. A consideration of geographical diversity, although limited in its scope, could be observed. Different imaginative geographies would coexist at the Palais du Trocadéro on 1st May 1878.

It should also be added that on the upper terrace of the round shaped main building were installed another thirty female statues representing "sciences, the arts, and technical applications." *La Géographie*, made by Maximilien Bourgeois (1839–1901), was among them (Exposition Universelle de 1878: 64–65). A model of *La Géographie* was donated to the Société de Géographie de Paris as "a precious souvenir of the 1878 Universal Exposition" (Girard 1878: 579). Internationalism, art, and geography were interrelated with each other in an intriguing manner at the Palais du Trocadéro in 1878.

8.3 Conclusion

This paper has dealt with the so far unexplored relationship between internationalism, art, and geography by discussing mainly the two cases of Paris, in the late nineteenth century. Among art, I focused here on the female statues representing the continents which had become conventional and popular in late nineteenth century Europe (Dupre 1991). In tackling this subject matter, I employed a couple of perspectives or approaches which seem relevant here. These include the historicity and spatiality of internationalism, the related concept of scale jumping, a landscape-as-text approach, and the related concept of imaginative geographies.

Late nineteenth century Paris was characterized as the site where occurred a host of projects and events promoting internationalism by means of some kind of scale jumping. Among others, the world's fairs and international congresses were typical events concerning internationalism, especially a hegemonic one. The erection of *Les Quatre Parties du Monde* was virtually linked to the debates and assertions surrounding the selection of the prime meridian, which was one of the themes at the two Paris international geographical congresses: the Congrès International des Sciences Géographiques in 1875, and the Congrès International de Géographie Commerciale in 1878. The physiognomies of *Les Quatre Parties du Monde* delivered an apparent message of the supremacy of Europe over Others, and the fountain and statues were intended to symbolize the Paris meridian itself and to assert its status as the starting point for the calculation of the world's longitude. On the other hand, the group of *Les Six Continents* was stemmed from the sort of hegemonic internationalism and imaginative geographies that found themselves at the 1878 Paris International Universal Exposition. While the appearances of *Les Six Continents* indicated a linear progress from primitive Oceania to civilized Europe, their arrangement at the inaugural ceremony at the Palais du Trocadéro signified the historical and geographical centrality of the Mediterranean world.

In concluding this paper, I would like to add some further comments. Firstly, it is generally accepted that the symbolism of women as part of the earth has a long tradition. It is also said that the practice of representing cities as female dates back to the tradition of the ancient Near Eastern mythology (Weems 1995: 44). And the representation of the continents through female figures dates back at least to the famous title page of *Theatrum Orbis Terrarum* published in 1570, by

Abraham Ortelius (1527–1598) (Shirley 2009: 46). Europe is depicted as an elegantly dressed woman on the upper side. Asia is depicted as an exotically dressed woman on the left side, and Africa is represented by a black, half-naked woman on the right side. Lastly, America and Magellanica are both depicted as a naked woman on the lower side. Therefore, it is evident that both *Les Quatre Parties du Monde* and *Les Six Continents* were the descendants of the title page of Ortelius' first world atlas. In addition to this, it could be said that the female statue was regarded as a more eye-catching, attractive medium than the male one in nineteenth century androcentric European society.

Secondly, in semiotic terms, the statues representing the continents could be understood as geographical entities in a double sense. On the one hand, as the signifier, they constitute part of the landscape of a particular place. On the other hand, they denote the notion of a particular continent (for example Africa) as the signified, where the referent is the African continent in the real world. Thus, the practice to erect the statues of the continents could be regarded as a geographical practice in a double sense. It is the practice to construct a small geographical world and it is also the practice to represent a large geographical world. An imaginary scale jumping is inherent in it.

Lastly, my assertion here is that the statues of the continents could also be understood as a different kind of "geo-body." The term geo-body was originally coined by historian Thongchai Winichakul. He stated that it "is a man-made territorial definition which creates effects – by classifying, communicating, and enforcement – on people, things, and relationships" and it "occupies a certain portion of the earth's surface which is objectively identifiable" (Winichakul 1994: 17). While his concept of geo-body refers principally to the territorial shape of a nation, which is depicted on the map, I would like to extend the scope of the concept to include diverse material artifacts representing geographical regions on the different scales. This extension of the concept of geo-body enables us to parallel the statue with the map in evoking some feelings, attitudes, and imaginative geographies. As the geo-body of the continents, both *Les Quatre Parties du Monde* and *Les Six Continents* would become "a source of pride, loyalty, love, passion, bias, hatred, reason, unreason" (Winichakul 1994: 17).

Acknowledgements This work was supported by JSPS KAKENHI Grant Number JP17K03248.

References

Alphand A (1867–1873) Les promenades de Paris. J. Rothschild, Paris

Anonymous (1872a) Compte-rendu du congrès des sciences géographiques, cosmographiques et commerciales tenu à Anvers du 14 au 22 août 1871. Tome premier. L. Gerrits et Guil. van Merlen, Anvers

Anonymous (1872b) Compte-rendu du congrès des sciences géographiques, cosmographiques et commerciales tenu à Anvers du 14 au 22 août 1871. Tome second. L. Gerrits et Guil. van Merlen, Anvers

Anonymous (1875) J. B. Carpeaux. The Architect 14:212

Anonymous (1878) Congrès international de géographie commerciale tenu à Paris, du 23 au 30 septembre 1878. Imprimerie Nationale, Paris

Araeen R (1994) New internationalism. Or the multiculturalism of global Bantustans. In: Fischer J (ed) Global visions. Toward a new internationalism in the visual arts. Kala Press, London, pp 3–11

Bellentani F (2016) Landscape as text. In: Higuera CJR, Bennett TJ (eds) Concepts for semiotics. University of Tartu Press, Tartu, pp 76–87

Benjamin W (1969) Paris. Capital of the nineteenth century. Perspecta 12:163–172

Bibliothèque de l'INHA (2002) Luxembourg, Fontaine Carpeaux. http://bibliotheque-numerique.inha.fr/idurl/1/4483. Accessed 7 Feb 2019

Bouthillier de Beaumont H (1881) Du choix d'un méridien initial unique. In: Anonymous (ed) Congrès international de géographie commerciale tenu à Paris, du 23 au 30 septembre 1878. Imprimerie Nationale, Paris, pp 211–213

Brockington G (2009) Introduction: Internationalism and the arts. In: Brockington G (ed) Internationalism and the arts in Britain and Europe at the fin de siècle. Peter Lang, Bern, pp 1–24

Bruzen de la Martinière AA (1768) Le grand dictionnaire géographique, historique et critique. Tome quatrième. Les Librairies Associés, Paris

Cortambert E (1879a) Choix d'un premier méridien. La Nature 7(300):199–200

Cortambert E (1879b) Selecting a first meridian. Popular Sci Monthly 15:156–159

Darlet A (1874) Fontaine de l'Avenue de l'Observatoire. L'Univers Illustré 17(1020):647–648

Desbiens C (2017) Imaginative geographies. In: Richardson D et al (eds) International encyclopedia of geography. People, the earth, environment and technology. John Wiley & Sons, New York, pp 1–6. https://doi.org/10.1002/9781118786352.wbieg0865

Destors D (1881) Notice sur la vie et les œuvres de Gabriel Davioud. Ducher, Paris

Duncan JS (1990) The city as text. The politics of landscape interpretation in the Kandyan Kingdom. Cambridge University Press, Cambridge

Dupre F (1991) La France et les autres. A study of colonial discourse in nineteenth century French public sculptures. Women's Art Magazine 39:8

Exposition Universelle de 1878 (1878) Le Palais du Trocadéro. A. Morel, Paris

Ferretti F (2016) Geographies of peace and the teaching of internationalism. Marie-Thérèse Maurette and Paul Dupuy in the Geneva International School (1924–1948). Trans Inst Br Geogr 41(4):570–584

Fratello B (2017) Rome or Babylon? The Gare du Nord in competing narratives of Second Empire Paris. In: Djokić V et al (eds) Entangled histories, multiple geographies. Papers from the international scientific thematic conference EAHN 2015 Belgrade. University of Belgrade, Faculty of Architecture, Belgrade, pp 92–97

Fuchs E (2002) The politics of the republic of learning: international scientific congresses in Europe, the Pacific Rim, and Latin America. In: Fuchs E, Stuchtey B (eds) Across cultural borders. Historiography in a global perspective. Rowman & Littlefield, Boulder, pp 205–244

Girard J (1878) Séance du 16 octobre 1878. Bulletin de la Société de Géographie 16:577–582

Grataloup C (2009) L'invention des continents: Comment l'Europe a découpé le monde. Larousse, Paris

Halliday F (1988) Three concepts of internationalism. Int Aff 64(2):187–198

Harvey D (2003) Paris, capital of modernity. Routledge, New York

Hodder J et al (2015) Introduction: historical geographies of internationalism, 1900–1950. Political Geogr 49:1–6

Jones CA, Clark J (2008) Nationalism and internationalism in modern art. In: Gopal S, Tikhvinsky SL (eds) History of humanity. Scientific and cultural development, vol VII: The twentieth century. Routledge, London, pp 408–446

Jones JP III et al (2017) Neil Smith's scale. Antipode 49 (1):138–152

Judd D (1992) Art and internationalism: Prolegomena. In: Judd Donald (ed) Shizuoka Prefectural Museum of Art. Gallery Yamaguchi, Osaka, pp 5–12

Keilo J (2015) The Gare du Nord and a Parisian view of Europe. https://centrici.hypotheses.org/620. Accessed 2 Aug 2017

Kish G (1989) International geographical union. A brief history. IGU Bull 39:110–119

Lyons FSL (1963) Internationalism in Europe, 1815–1914. A. W. Sythoff, Leyden

Margerie LD (2014) Fountain of the Observatory. In: Draper JD, Papet E (eds) The passions of Jean-Baptiste Carpeaux. The Metropolitan Museum of Art, New York, pp 156–167

Muller H (1867) Le pavillon de Flore. L'Univers Illustré 10(631):227

Otlet P (1911) Le musée international et les sciences géographiques. Bulletin de la Société Royale Belge de Géographie 35(6):452–458

Pingeot A (1994) La sculpture du XIXe siècle. La dernière décennie. Revue de l'Art 104:5–7

Rasmussen A (1989) Les congrès internationaux liés aux expositions universelles de Paris (1867–1900). Mil Neuf Cent 7:23–44

Rice CE (2016) Architecture's capital? Revisiting Benjamin's Paris. In: Clayson H, Dombrowski A (eds) Is Paris still the capital of the nineteenth century? Essays on art and modernity, 1850–1900. Routledge, London and New York, pp 19–34

Shimazu T (2015) War, peace, and a geographical internationalism. The 1871 Antwerp international geographical congress. Geographical Reports of Tokyo Metropolitan University 50:97–105

Shirley RW (2009) Courtiers and cannibals, angels and amazons. The art of the decorative cartographic titlepage. Hes & De Graaf, Houten

Société de Géographie (ed) (1878) Congrès international des sciences géographiques tenu à Paris du 1er au 11 août 1875: Compte rendu des séances. Tome premier. E. Martinet, Paris

Société de Géographie (ed) (1880) Congrès international des sciences géographiques tenu à Paris du 1er au 11 août 1875: Compte rendu des séances. Tome second. E. Martinet, Paris

Sterling and Francine Clark Art Institute (2011): Eye to eye: European portraits 1450–1850. https://www.clarkart.edu/exhibitions/eyetoeye/content/visitor.cfm. Accessed 10 Feb 2019

Tapia C, Taieb J (1976) Conférences et congrès internationaux de 1815 à 1913. Relations Internationales 5:11–35

Weems RJ (1995) Battered love. Marriage, sex, and violence in the Hebrew prophets. Fortress Press, Minneapolis

Winichakul T (1994) Siam mapped. A history of the geo-body of a nation. University of Hawaii Press, Honolulu

Withers CWJ (2017) Zero degrees. Geographies of the prime meridian. Harvard University Press, Cambridge, MA

Pierre Monbeig and the Geohistory of Brazil

9

Larissa Alves de Lira

Observe the geographical history that we rehearse by promoting and baptizing it as geohistory (Braudel 1951: 486).

Abstract

A long-established research theme in the realm of the intellectual history of the human sciences in Brazil, consists of identifying the different ways in which national scientific communities were shaped by French mentalities, since the Faculty of Philosophy, Sciences, and Letters of the University of São Paulo, the first Brazilian university, was founded by a French international mission, sent to Brazil, in 1934. However, the direction of influence can also be seen in plural perspectives as a subject of research, and by trying to reach a complex combination about how a different manner of thinking about the formation of capitalism in a peripheral country was also produced in Brazil. How did the Brazilian context and Brazil's territory help to shape French geographical epistemology? Rooted in the influence of Vidal de la Blache, Fernand Braudel and Brazil's territory (as a space of conception), my major hypothesis is that the young French geographer Pierre Monbeig, conceived an original geohistory of Brazil, in its principles, in parallel, and in harmony with the geohistory of the Mediterranean world.

Résumé

Un thème de recherche établi depuis longtemps dans le domaine de l'histoire intellectuelle des sciences humaines au Brésil consiste à identifier les différentes manières dont les communautés scientifiques brésiliennes se sont formées à partir des influences françaises au sein de la Faculté de philosophie, sciences et lettres de l'Université de São Paulo. Celle-ci, la première université brésilienne, a été fondée par une mission française envoyée au Brésil en 1934. Toutefois, les influences peuvent également être envisagées dans une perspective plurielle avec des rétroactions vers la France à partir d'une manière différente de penser la formation du capitalisme dans un pays périphérique. Comment le contexte brésilien et le territoire

This article was produced from research that obtained financial support from the Foundation for Research Support of the State of São Paulo (FAPESP).

L. A. de Lira (✉)
Federal University of Minas Gerais (UFMG), Belo Horizonte, Brazil
e-mail: lara.lira@gmail.com

B. Schelhaas et al. (eds.), *Decolonising and Internationalising Geography*, Historical Geography and Geosciences, https://doi.org/10.1007/978-3-030-49516-9_9

brésilien ont également contribué à façonner la théorie géographique française? Mon hypothèse majeure est que, sous l'influence de Vidal de la Blache, de Fernand Braudel et imprégné de sa connaissance du territoire brésilien, le jeune géographe français Pierre Monbeig a conçu une approche géohistorique originale dans ses principes, en parallèle et en harmonie avec la géohistoire du monde méditerranéen.

Resumen

Un antiguo tema de investigación en el ámbito de la historia intelectual de las ciencias humanas en Brasil es la identificación de las diferentes formas en que las comunidades científicas nacionales fueron inspiradas por las mentalidades francesas desde cuando la Facultad de Filosofía, Ciencias y Letras de la Universidad de São Paulo, la primera universidad brasileña, fue fundada por una misión internacional francesa enviada en Brasil en 1934. Sin embargo, la dirección de las influencias puede también verse en perspectivas plurales como un tema de investigación, y como tentativo de alcanzar una combinación compleja para reflexionar a como una manera diferente de pensar en la formación del capitalismo en un país periférico también se produjo en Brasil. ¿Cómo ayudó el contexto brasileño y el territorio de Brasil a dar forma a la epistemología geográfica francesa? Enraizado en la influencia de Vidal de la Blache, Fernand Braudel y el territorio del Brasil, mi asunto principal es que el joven geógrafo francés Pierre Monbeig concibió una geohistoria original de Brasil en sus principios, en paralelo y en armonía con La geohistoria del mundo mediterráneo.

Keywords

Geo-history of Brazil · Pierre Monbeig · Vidal de la Blache · Fernand Braudel · Semi-periphery · Actors and subjectivity

9.1 Introduction

The Mediterranean and the Mediterranean World, in the Age of Philip II (first edition, 1949, second edition, 1969), is a book that inaugurated a new field of research in the human sciences: geohistory.[1] In this field, the concept of long duration prevails over other rhythms and historical temporalities (conjectures and events) of history. These temporalities altogether form what Braudel called the "dialect of duration." The concept of "long duration" is based on a critique of traditional history, the large-scale reading of documents, the fertility of the ideas of the French geographer Vidal de la Blache (Lira 2008), and the consideration of space dynamics that Braudel, witnessed in fieldwork in Africa and Brazil. Considering all these factors, Braudel, used "geographic time" as a synonym of long duration.

Originally used by geologists to understand deep history (Rudwick 2014), geohistory brought to the human sciences the idea that space produces a frame for gradual evolutions. The link between geology and the human sciences was especially made by Vidal de la Blache,[2] who formulated that an evolutionary history has different stages according to its dislocation in space.[3] Fernand Braudel, played a special role in gathering these principles, identifying them as a new field of research. However, the principles had already been formulated by Vidal de la

[1]Braudel's geohistory is part of the Annales School, normally understood from an intellectual history approach by putting accent in ideas but also in institutional and relations of power in academy. See Burke (1997), Daix (1999), Dosse (2004[1998], 1992[1987]), Gemelli (1995), Lacoste (1989), Reis (2000), Romano (1997[1995]).

[2]For more recent publications about Vidal de la Blache, see Robic (2000, 2004), Clout (2003), Courtot (2007), Bord et al. (2009), Lira (2013), Haesbaert et al. (2012). These works put evidence in the plurality and actuality of Vidal de la Blache in sensitive objects of the French Geography, such as the problems of State, globalization, geopolitics, etc.

[3]Which is also, probably, a Darwinian principle. See Geraldino (2016).

Blache, as Braudel recognized.[4] Vidal de la Blache wrote:

> I would add that, from this point of view, the whole order of new relations opens up to the spirit. Because the *action of time* enters as the most important coefficient in the actions of natural causes. As *the zones* are more or less advanced in their evolution, they go through a series of changes that are linked to *each other by a kind of affiliation*. Some still retain traces that have already been abolished in others. It is as if they were living examples of the same phenomena taken in various stages. (Vidal de la Blache 1896: 146) [italics added]

This paper analyzes how the legacy of Vidal de la Blache, in regard to geohistory, was transformed, from a theoretical point of view, by the encounter of Pierre Monbeig[5], with the Brazilian countryside,[6] elaborating on theoretical contributions that were original but also in complement to Braudel's geohistory, as well as contemporary. Therefore, in addition to the influence of Vidal de la Blache on Monbeig,[7] I also have to consider the role of Brazilian spaces and their importance for Monbeig's elaboration. With this in mind, in some opportunities, I compare Monbeig's movement to Braudel's, considered the most important scholar in the domain of geohistory.

Both Fernand Braudel and Pierre Monbeig were part of a French scientific diplomatic mission sent to the University of São Paulo, in 1935,[8] one year after the foundation of the first Brazilian university (USP). Fernand Braudel brought to Brazil, his documents related to his visits to Mediterranean ports and cities, based on which he began to write his thesis, which became his masterpiece, *The Mediterranean and the Mediterranean World, in the Time of Philip II*, while Pierre Monbeig and Claude Lévi-Strauss, another member of the French mission, worked on the Brazilian terrain (Loyer 2015).

My theoretical approach seeks to associate the characteristics of the Brazilian *space* with geography's *epistemology*. This reflexivity was previously established by Dereck Gregory, who followed the innovation of the historiographical tendency known as "spatial turn." He showed that the *configuration* of travelers' thinking in the nineteenth century was related to the *configuration* of the space in which they traveled.

> Three topologies are identified that entered into spatial formations of knowledge in the long nineteenth century. The first is a *rhizomatic space* (a complex space with no center), illustrated through crucial readings of Alexander von Humboldt's travels through *the Amazon* and Mary Kingsley's travels in West Africa; the second is a *labyrinthine space* (a complex space with a center), illustrated through Sophia Poole's journeys through *Cairo*; the third is *a striated space* (an ordered, *linear space with a center*), illustrated through the ways in which *Europe* was itself scripted by the Grand Tour and the modern guidebook. (Gregory 2000: 297)[9] [italics added]

From my point of view, Pierre Monbeig elaborated in Brazil a systemic, open, expansive, and *active* geohistory, considering the geographical *configuration* of Brazilian space as a pioneer fringe: an open and vast territory, where with the advent of the twentieth century men were transforming spaces. For that matter, my results point to the following conclusions. Similarly to Vidal de la Blache, Pierre Monbeig thought about the Brazilian space based on the concept of *genre de vie*. Similarly to Braudel, he put the characteristics of the Brazilian space before all other historical movements and developed an expansive and systemic geohistory.

[4]"One of the most fruitful works of history, perhaps even the most fruitful of all, will have been that of Vidal de La Blache, a historian in origin, a geographer in vocation" (Braudel 1992: 50). The author has translated all quotes into English, except Gregory 2000 (p. 297).

[5]Pierre Monbeig is not well known out of the Brazilian community of geographers and the little world of French historians of geography. Classical works about him include Théry and Droulers (1991) and Salgueiro (2006a). More current works include Clout 2013 and Lira, book in press.

[6]I see the territory of Brazil in a specific context and as a space of conception for geography as science.

[7]Pierre Monbeig can be considered a member of the French Geographical School founded by Vidal de la Blache.

[8]For more details about the French Mission, see Massi (1981), Mota (1981), Limongi (1988), Peixoto (2001, 2008), Petitjean (1991, 1996a, b, 1998), Suppo (2000, 2002).

[9]About a geography of science, see also Livingstone (2003) and Whithers (2007).

Fig. 9.1 Braudel in a little restaurant in the countryside of Brazil. This and the other figures show how they make geography and history in Brazil in this moment by exploring the inner land. *Source* Lévi-Strauss (1994)

However, in slight contrast to Braudel, Monbeig focused on the subjectivity of actors, which was one of his originalities. In this way, he conceived of a systemic, open, expansive *and active* geohistory, because Brazil in the twentieth century was a frontier, like the Mediterranean Sea in the sixteenth century, but with more active actors, who lived in the twentieth century. Thus, this displacement of space and time is also taken into account in understanding geographical thought (Figs. 9.1 and 9.2).

Fig. 9.2 The young Pierre Monbeig. Pierre Monbeig came to Brazil when he was 27 years old. He was very enthusiastic for exploration the countryside. *Source* Booklet commemorating the Pierre Monbeig Chair. FFLCH-USP

9.2 The Role of Vidal de la Blache for the General Conception of Geohistory

Did Geohistory exist in the human sciences before Braudel? Vidal de la Blache received a bachelor's degree in history. He charted his institutional trajectory according to the possibilities that emerged at the end of the nineteenth century in France (Andrews 1986). He always maintained a dialog with the university's historians. During this institutional process, Vidal de la Blache, the historian was becoming a geographer through slow, institutional transitions, and successive ruptures (Lira 2013). In this gradual process, a concept of time was becoming a concept of space. In the paper *Sur l'esprit géographique* he defined the necessity of tracing "new perspectives of time": "We must step away from the past, get used to other perspectives of time. The geographer's clock is not exactly the same as that of the historian" (Vidal de la Blache 1914: 557).

Vidal thought about a concept of time that could be observed in the landscape, focusing on crops' expansion, diffusion, and the uses of farming instruments. He was also an exceptional reader of archives, linking the "geographical point of view" with his reading of sources. Supported by historiography, he was able to see in the landscape a time that materialized in space. In the Mediterranean of Vidal de la

Blache,[10] I can see successive circles of used lands, a forward and backward movement in the colonization of space, ebbs, and flows of technique, roots of genre de vie, a dialectic of modern and primitive land use according to crisis periods —men and techniques in association and in movement forming a chronology of human action in contact with nature. Vidal de la Blache was not concerned with calling attention to geohistory because for him all of geography was a kind of history.

Over the years, Braudel intended to expand history over other subjects and themes. At the time of writing his Mediterranean history, he defended the view that as important as it was to understand the men, exchanges, and trade in this economy, it was essential to look to a wider space, and to how the English tried to create roots that diverged from the Mediterranean. In the marginal spaces of this economy, it was building future history (Braudel 1983[1969]).

Braudel took from Vidal de la Blache the intuition of scaling up to observe evolutions. Braudel took this idea to demonstrate that only a global history could explain the history of the world in its different temporalities.[11] The history of an isolated center would induce a teleological history, while a history including the marginal spaces was a geohistory, which is systemic. The amplifying of the Mediterranean Sea stemming from the maneuvering of the English merchants at its margins is one of the most important processes in the formation of pre-industrial European history.

Now, how do these facts concern Brazil? What were the Brazilian characteristics of space noticed by Monbeig that were reflected in his thinking? In fact, I can see a kind of reflection between space as an object of research, and the epistemological aspects of the thought of Fernand Braudel and Pierre Monbeig, in the Mediterranean World but also in Brazil, which is a legacy of the latter with regard to French geography, that is not well known.

9.3 The Role of the Spaces for a General Conception of Geohistory

Could Brazil, be like the Mediterranean? Like the Venetian, or Genoese external trade maneuvering around from the heart of the Turkish Empire, or the English merchants maneuvering around the Mediterranean Sea, through the Baltic, showing that commercial relations developed around old economies and political powers, Brazil, was also a marginal space in the European world economy at the beginning of the twentieth century. In Brazil, Pierre Monbeig started to research the pioneer fringes of São Paulo and Northern Parana. The pioneer fringes were areas of recent modernization. They were associated with a more global transformation of capitalism, as well as a process in the Brazilian economy of transition, from a poorly productive system to a more highly productive system, pushed up by the margins. This process was paradigmatically analyzed by the Belgian historian, Henri Pirenne, who studied the transition from feudalism to capitalism in Europe, in the margins of this area (Pirenne 1963[1933]). The capital of the political power in Brazil was the axis between Rio de Janeiro and Minas Gerais.

Monbeig made little effort to situate the Brazilian case amongst other paradigmatic processes, although it was. The evolution of the pioneer fringes in Brazil is more a paradigmatic case born of capitalism in a marginal position from the political power or the old economies. The comparison between Brazil and medieval Europe was explicitly made by Monbeig.

> The region's mother highway, the urban foundation, which thrives thanks to the protection of a powerful man, the land's delimitation, an initially disorganized exploration of the land that little by little tends to organize itself better, none of this seems so original when you think about the geography of European campaigns. Speaking of the High Middle Ages, Marc Bloch describes the agglomerations, in which men lived close to one

[10]Vidal de la Blache also produced important papers about the Mediterranean. See Lira (2013).

[11]Vidal de la Blache also formulated the earth's unity principle. See Vidal de la Blache (1896).

> other. But these agglomerations were separated by multiple voids: yields of the crop systems were weak, and large tracts of arable land were needed: the fields gained through temporary and brief conquests did not pass over uncultivated lands… incessantly tending to overlap nature. And further on, a description of *the tropical farm* resembles this analysis of the *landlord*: it comprises two areas, on the one hand, the realm whose fruits the master collects directly; on the other, the concessions and small or medium farms. These concessions were formed by distributing lands to the slaves [to be cultivated], transformed into tenants [the landlords], at a time when the vast spaces of the plantation became poorly profitable through direct exploitation. (…). Is it not the same evolution? Mutatis mutandis; history does not repeat itself. (Monbeig 1984[1952]: 389) [italics added]

In consequence, as in the Middle Ages, the study of pioneer fringes pointed to a scientific problem linked to the development and growth of the economy. However, in contrast to the Mediterranean World and medieval European history, it was a development framed by Brazil's own time and space.

9.4 The Role of Brazilian Context for Monbeig's Conception of Geohistory

Space was not solely a frame for Pierre Monbeig's conception of geohistory. From my point of view, the Brazilian space also structured the political, economic, cultural, and intellectual context, in which Pierre Monbeig was included. An extensive debate showed a gradual growth of awareness regarding the dimension of Brazilian space. Since the independence of Brazil, from Portugal in 1822, important politicians began to discuss the necessity of moving the capital from the coast to the interior to start to colonize all the territory. In 1848, the Belgian astronomer and geologist Louis Crouls, carried out a large number of surveys in the Central Plateau, to try to define the location of the national capital. In 1922, the Army made its choice and produced a map where the federal district is placed in the middle of Brazil.

In the thirties, when Pierre Monbeig arrived in Brazil, the territory began to be an important subject of debate for scholars. We reach the end of this decade with an important simultaneous progression in the growth of the scientific community, the emergence of the question of development and the growth of awareness regarding the scale of the Brazilian space. The movement of the "March to the west,"[12] carried out personally by the President, Getúlio Vargas, was also: "(…) the time of young revolution; an installation of modernity". A real consciousness of a national reality begins to appear in the cultural framework of Brazil, starting with the denunciation, first off, of the state of underdevelopment of the country (Getúlio Vargas, public domain, without page).

The federal institutions and democracy itself shaped themselves to accomplish substantial social and territorial plans. The Army claimed autonomy from Congress, and Vargas instituted the New State, an authoritarian and dictatorial state with the excuse that democracy in Brazil, was not capable of solving extensive, persistent, and large territorial problems (Neto 2013). Getúlio Vargas spoke at the inauguration of the New State:

> We need to equip the country's railways to offer economic transport to the products of *different regions*… On the other hand, these achievements demand that a large steelworks industry be created… There is an urgent need also to equip the armed forces with sufficient equipment (…) *In order to readjust the political organism to the economic needs of the country* and guarantee the measures indicated. No alternative was offered other than that which was taken, establishing a *strong regime* of peace, justice and work. (Getúlio Vargas, public domain, without page) [italics added]

In the Army, the relation between the territory and the institution's needs was even closer:

> The vastness of our territory, about 8,061,260 square kilometers; the very extensive frontier

[12]Ricardo Cassiano wrote the book "March to the West" in 1940, and other authors, such as Mario Travassos, also contributed to formulate a public opinion about the need to occupy the countryside as the solution for Brazilian social problems (Cassiano 1970 [1940], Travassos 1938).

> boundary that circumscribes it; the difficulty that still exists regarding the means of communication; the density of the population, which though small in most of the national centers considered as populous, is almost null, or null, in a large portion of the country; (…) are elements that disturb and harden the problem of our defense and show that its resolution can only be obtained by the establishment of *a broad and wide project, whose execution is carried out in a consistent, systematic and continuous way*. In no other manner can the great undertakings be realized. The geography of the country and its topographical configuration indicate that the establishment of this project, requiring the simultaneous use of sea and land forces, must be done *by a National Defense Commission composed of admirals and generals and submitted to the examination and study of the Congress*, who will transform it into law, the execution of which will be resolutely and invariably carried out, nevertheless, through periodical governmental successions. (Martins 2017: 166) [italics added]

In the domain of culture, an important writer announced in one of the period's most important books that: "One of my interests was to disrespect the geography and the geographic fauna and flora. [But also] (…) considering Brazil, a homogeneous entity, a national and geographical ethnic concept" (Preface by Mário Andrade to Macunaíma. In Coutinho 2017: 95).[13] All this context had an effect in Pierre Monbeig's awakening to some debates: "Being in Brazil, Monbeig is aware of the challenges facing geography *earlier than* if it remained in Europe. He is sensitive to the demand for development that presents itself for Brazil, of the New State" (Dantas 2005: 18) [italics added].

9.5 The Role of the Institutional Context for Monbeig's Conception of Geohistory

The foundation of the University of São Paulo in 1934, where Pierre Monbeig and Fernand Braudel taught in Brazil, in this context could also be interpreted as a dispute between how to carry out development in a *capitalistic way* or in a *"territorialist"* way.[14] While Getúlio Vargas clearly favored a *"terrorialist"* way of development, the *paulistas*, inhabitants of São Paulo, preferred a *capitalist* way of development. Thus, Getúlio Vargas tried to establish a populist state by controlling its continental roots and moving the capital to the center of Brazil, while the *paulistas*, a liberal regional group led by Júlio de Mesquita Neto (he was also a founding father of the University), wished to draw the internal roots toward an external and maritime exit in São Paulo. In 1932, the "Revolution of 1932" broke out when they tried to tear power away from Getúlio. Having lost the internal war, the dispute migrated to culture. For the *paulistas*, it was urgent to spread "democratic" values across the territory. Even if employed by the *paulistas*, Pierre Monbeig, Fernand Braudel, and Claude Lévi-Strauss formed a kind of dissident group and tried to guarantee their teaching autonomy from the University (Fig. 9.3).

> And that in the vision of the founders of the University of São Paulo, the Faculty of Philosophy, Sciences and Letters would unite all branches of science in one institution, giving it a high capacity for generation and preservation of a new mentality. The *Paulistas* were in a hurry. They could not waste time or resources. Júlio de Mesquita Filho was aware of this, well before the Revolution of 32. The investigation showed him that in order to build the University, he had to completely reform teaching: primary, secondary, and middle school, as well as the university. Then he said, 'We have to do all of this. Now, if we start with primary school, logically, it should prepare students for secondary school, secondary school should prepare students for the university, and it will take ten or twelve years. Let's do the opposite. Let's found a University and, inside the University, we create a model gym. ' It would be a high school like in France. And in the university, we prepare to teach in the gymnasium. (Duarte, interview.) And it was with great haste that Mesquita set to work, soon after returning from exile, to build the University. (Schwartzman 1979: 200)

[13]"Geographic" in this context appears in a "regional" sense and the sense of "unity" of the country.

[14]The capitalist way of development involves a multipolar political order subject to economic power. In contrast, the territorialist way involves a single political power that controls economic power (Arrighi 2007).

Fig. 9.3 Dina Lévi-Strauss, Fernand Braudel, Pierre Monbeig and Jean Maügüê exploring the inner land. Also, an example of the making of geography and other human sciences in Brazil. *Source* Lévi-Strauss (1994)

In this movement to guarantee their autonomy, they made an important alliance with the most underprivileged sector of the University of São Paulo: the students. Conflicts started to arise in São Paulo between the professors and the employees, in which, student support was very important:

> On September 17, philosophy students took the initiative to offer Professor Maügue a Tea of Honor. All the French teachers participated. (…) On this occasion, the students publicly expressed their appreciation for the work of the French professors and especially for the spirit of intellectual freedom present in how they conducted their work. (Archives diplomatiques (France)—La Courneuve —box 444—D series, carton 170, dossier 9, *Informations sur la mission universitaire de Sao Paulo [juillet, sept 1938]*)

The University of São Paulo developed a more historical than practical point of view for understanding Brazil. The scholars in São Paulo, based on the French model, were not directly focused on the question of practical's geographies. History and Philosophy were the keys points of this model. In the following topics, I will try to relate the *configuration* of the Brazilian space (as Pierre Monbeig conceived it) to a systemic, open, expansive, and *active* geohistory (Fig. 9.4).

9.6 Tropicality, Rhythm of Development and a Superstructural Dynamism of Capitalism in Brazil

One of these students from the University of São Paulo had an important influence on the young Pierre Monbeig, who was the same age as him. The relation between tropicality and economy has already been suggested by this communist and historian, Caio Prado Jr.[15] According to him, because of its tropicality, the "new country" was totally different from the United States, due to the immediatist greed with which the Portuguese, has taken control of the Brazilian environment. Because of an opposition to and compatibility with the temperate environment, products such as redwood, sugar, and coffee has been traded early (Iumatti et al. 2008; Pericás 2016).

[15]This sense of tropicality of Caio Prado Jr. does reference Vidal de la Blache, mainly, and perhaps Pierre Gourou, by searching different geographical information in the labor of the Associação dos Geógrafos Brasileiros (Iumatti et al. 2008).

Fig. 9.4 Pierre Monbeig and his students. *Source* CAPH, Center for Research in History, University of São Paulo

The Brazilian tropicality also produced a semi-primitive *genre de vie*.[16] In Brazil, I can view the prominence of a semi-sedentary agriculture characterized by the practice of slash-and-burn as an inheritance from the natives (Monbeig 1985[1954]). The implication of this was a low degree of density and wealth accumulation, as well as low labor productivity. Because of nomadism, agricultural technology was underdeveloped, the landless divided, the landscape not humanized, and disease resistance was weaker.

Thus far, all these facts were known in the historiography and incorporated by Monbeig. What Monbeig showed, however, in an original point of view, is that the capital that was expanding in the tropical environment did not construct settlements nor compensate for the low labor productivity. On the contrary, it took advantage of workers' mobility to devalue the price of labor in the market. In Brazil, the main internal association, the *plantation*, perpetuated the roaming of the population. Capitalism in Brazil does go deep in the transformation of the old economy of the country and remains superstructural.

> It is perhaps necessary to remember the financial needs of farmers and the organization of agricultural credits used to subsidize these needs. Rarely is there any other option than to address the capitalists, whose demands are disastrous. The only thing left to the farmer is a solution that gives him courage: the illusion that moving to another place, leaving one place for another, will improve things. When one learns that in some regions, which are the poorest and least technically developed in São Paulo, all the crops are bought by a single trader and all credit depends on that same trader, the situation is better understood. It is a ray of light projecting onto the landscape. (Monbeig 1957: 217)

According to Monbeig, the pioneer fringe was a space of mobility and exploration at its front, and only residually of settlement at its periphery (see Map 9.1). The dynamism was at its front where the *plantation* was dominant. Only as a subsidiary effect can I see the small farming and industrial cycle, attaching men to the land in small farms. Capital modifies the *genre de vie* only at its periphery. Capitalism in a tropical situation does not develop the land as its main process. For Monbeig, did tropicality relate to underdevelopment?

To Monbeig, the development would be measured in terms of rhythms, and here I return to the subject of geohistory. The dynamics that produce a delay are not without major consequences. This

[16]*Genre de vie* is a concept from Vidal de la Blache (1911).

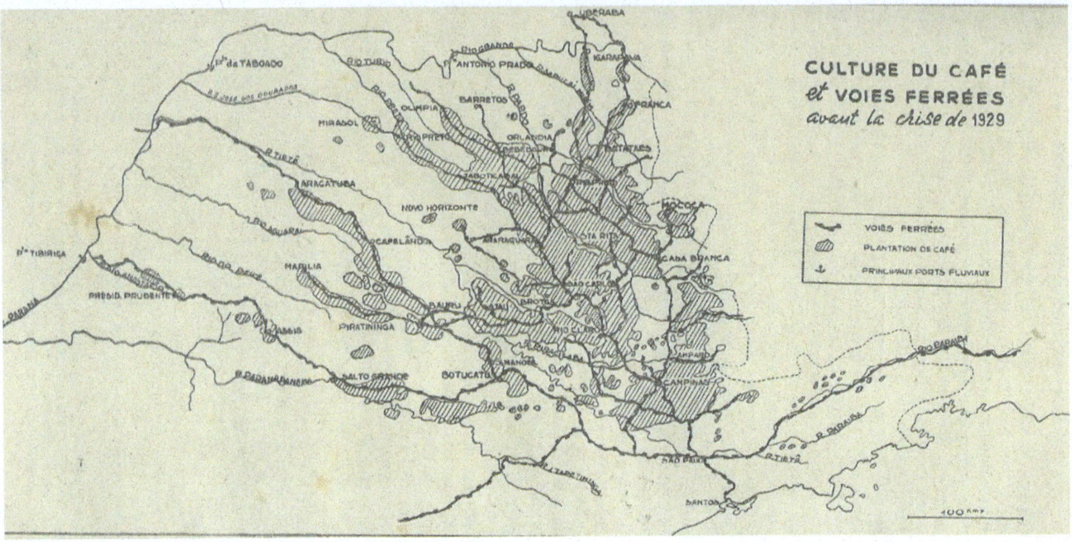

Map 9.1. Coffee production. *Source* Monbeig (1952)

delay is very important to economic growth in the nineteenth century. I argue that it is this delay in developing the land, identified by Monbeig, that explains (along with other reasons) the divergence between both Americas in the nineteenth century and that helps to explain the emergence of underdevelopment in the age of revolutions, when the Industrial Revolution spread across the world (Hobsbawm 1996).[17]

9.7 The Role of the Local Actors Related to the Geographical Vocation to a Semi-periphery

However, even if Europe had a high expectancy of colonial profits, the distance between Brazil and Europe helped to relax the pressure of imperial power. In the map of Vidal de la Blache in the nineteenth century, Brazil is a deviating root in the network of West-East trade (Vidal de la Blache 1894[1981]). The market's advance in a North-South direction was slower because the latitude gradient changes the climate.

Among the tropical zones, Brazil was further from the center of the capitalist world economy than Africa.[18] Indeed, the distance was based on the relative freedom enjoyed by colonial elites. They organized the state by themselves to fight against or incorporate the natives. Indeed, Brazil's colonization was also pushed ahead by internal forces. In the nineteenth century, it emerged as a semi-peripheral country in the capitalist world economy advertising modernity at the Paris World Exhibition (Lira and Clerc 2018).

Monbeig perceived that colonial elites succeeded in accumulating capital and in being the leaders of the organization of the state. Colonial elites benefitted from exchanging favors with the Portuguese crown (which moved to Rio de Janeiro at the beginning of the nineteenth century). After the dissolution of the empire, these elites "captured" the state and organized the advancement of capitalism in the Brazilian territory.

[17]A student of Monbeig, Nice Lecoq Müller, calculated a cycle of 40–50 years before the peasant would no longer be exploited in the countryside and become a farmer on the outskirts of the metropolis (Müller 1951).

[18]There is a generalization in this notion of "Africa" as a unique space. However, I d't want to compare the whole process of national capitalists, but the importance of distance between the European continent, that would interfere in this formation.

This relative freedom was noted by Pierre Monbeig, upon reflecting on culture and mentalities when he talks about a "bandeirante" mentality. The "bandeiras," Monbeig explains, are the "expeditions that, in colonial times, set out to explore the interior of Brazil. The "bandeirantes" were the members of these expeditions" (Monbeig 1952: 11). Some years later, the American brazilianist Nathanael Leff (1991), demonstrated that the Brazilian elites managed to organize their economic policies based on local interests. Leff's interpretation agreed with Monbeig's, in terms of having as a prerogative the attribution of an important degree of freedom to human action, even if these actors were placed in a peripheral space of capitalism in general terms.

Monbeig's geohistory is slightly different from Braudel's, as he posits a larger margin of maneuver for individual action and focuses on a region that moved from the periphery to a semi-peripheral situation in the twentieth century. Braudel dubbed Philip II a "paper-maker" to point to the lack of entrepreneurship of the Spanish King. Meanwhile, Pierre Monbeig put forward the idea that the landowners developed a "bandeirante" mentality as a kind of propaganda to attract the workers to seek "adventure" at the pioneer fringe. Therefore, these ideas contained an original point that emerges from the experience with the Brazilian space. By studying the formation of *late* capitalism, he could not put aside the active role of actors (Roncayolo 2006).

However, certainly not all of it can be attributed to the human capacity for producing space. According to Monbeig, Brazil continued to be a victim of natural handicaps that acted as an economic delay. One of these handicaps was a territorial structure that made circulation difficult. Developing local infrastructure was an old challenge for Brazilian geography, which increased over centuries of plans and projects (Sousa Neto 2012). Émile Levasseur and Vidal de la Blache both referred to Brazil as a tropical and "exuberant" country to emphasize this "closed" characteristic of the local environment resisting settlement. Fernand Braudel talks about the rhythms of extremes, according to Lima:

> The geography of the Brazilian territory is summarized as follows: the "tabular" Brazil, schematically, is cut in two by an abrupt drop of almost one thousand meters. On one side, a coastal platform barely separated from the waters; on the other, a huge plateau that descends gradually towards the Andes. The big problem is moving from one to the other, 'from the curb to the platform.' [...] Another problem arises for Braudel: the North and the South of Brazil are also different at this point. Their two stories are not synchronized. It is necessary to be attentive to the rhythm of these two extremes. (Lima 2009: 170)

In addition to this exuberance, nearly every point in the territory is far from every other, and the overly vast regions raised difficulties in terms of the "evolution" of natives or of Portuguese colonization. The native *genre de vie* was delayed before the immensity of the territory, and the "whites" would also rather have scattered across the South than deal with the forest (James 1950).

9.8 The Brazilian Space is Opened Over the World-System: The Importance of the Economic Cycles

On the whole, Monbeig's argument is that Brazil is an intermediate "player" in the global geohistory, neither very close nor very distant from the dynamic center of capitalism. In the Americas, it cannot be considered a dependent economy like the Caribbean nor independent like the United States of America, which continentalizes its insular dynamics (Arrighi 2007).

While Monbeig recognized this relative autonomy, he also noticed that Brazil was condemned to recover its deficit of capital in the global market.

Thus, economic development occurred for both internal and external reasons. Oligarchic elites attached themselves to global forces as a remedy to this economic delay, making the country an extension of the global market, acting as an island. Territorialism searched, however, for the labor forces scattered throughout the continental territory. Nevertheless, in the era of revolutions, during the First Brazilian Republic (1899–1930), territorialism was a minor force, which also coincided with the coffee boom, the main concern for Monbeig.[19]

In this way, the price variation of coffee in the global market was hugely important to Brazilians: the colonization of the *plateau* was driven by global markets through which Brazil could try to recover its deficit. However, this solution could not fail to pose some problems:

> Thus, the dynamism of the São Paulo settlement front is confused with that of the world economy. (…) The pioneer fringe is the meeting point of the appetites and ambitions of both nationals and foreigners. This exposes her to being hard hit by crises, but at the same time allows her to recover her vitality very quickly. (Monbeig 1984[1952]: 118–119)

On the basis of a solid geographical reality, Brazilian history is cyclical because it is determined by capitalist dynamics. However, the cycles are not only economic. Brazil's geographical reality is also shaped by a cycle of depletion of its tropical soils. The economic and pedological cycles combine to drive the colonization of vast lands in a frenetic search for environmental resources. In the coffee boom, capitalist dynamics took advantage of "territorialism," which had consequences in Monbeig's thought.

One final main characteristic of Brazil, noted by Monbeig, is that it was a frontier space, where the economic process was in movement. With that, Pierre Monbeig adapted the geohistory logic to understand an internationalized (Salgueiro 2006b, c) and dynamic reality where the actors have pushed ahead in an open and expansive system.

9.9 Monbeig's Originality: An *Active* Geohistory, with Systemic and Expansive Vision

I argue that systemic thinking has its origins in a naturalistic vision of the history of the earth, which is why geography is so important as a link between the natural and human sciences. In 1883, Vidal de la Blache published a school book showing that the circular spatiality in geography first came from advancements in physical geography (Azevedo 1976). Nonetheless, according to Daniel Loi, the adaptation of systemic thinking was, until the Second World War, unused or at least in a latent state (Loi 1982, 1985). It is possible that it was in Brazil, that the pioneer application of a systemic geohistory in empirical research on human geography took place, in parallel to the history of the Mediterranean Sea, which was open, expansive and, in addition, *active*. Braudel's thinking was also systemic, but it refers to spaces produced by human action before the Industrial Revolution and that were very *passive* from the point of view of the actors on stage.

In the beginning of the twentieth century, Brazil was one of the few countries that were still a frontier. In these terms, it was a special case to global geohistory: in Brazil, we had observable forms of capitalist expansion. Brazil, being a vast country with exploitable lands pushed Monbeig to try to understand why capitalism expanded through the territory.

He conceived of a structural instability inside the capitalist circuit preventing it from closing its movement at the same point where it started in space. Imagining it functioning as a circuit, it should have an internal instability that pushes it to expand through space. For him, this internal instability had as its *leitmotiv* the association with the exhaustion of the soil, with the distance from the center of European capitalism, with the increase in prices in the global market, with the abundance of available land *and*, which is very important, with *the conservative mentality of elites in power* that tried to increase labor

[19] For the definition of Arrighi (2007) of capitalism and territorialism that I adopted, different logics involving multipolar economic order or a single political power can be added. Based on this classification, I believe also that Arrighi also highlighted how these different logics were combined.

mobility instead of developing it on the soil. Thus, capitalism always found further advantages. I argue that Pierre Monbeig's thinking demonstrated a pattern of modernization in the New World based on mentalities.

However, for Monbeig the origin of capitalism and its advance through space come not only from a systemic movement. Even with clear structural determination, there are also individual choices. The success and failure of individuals lead to rapid urban growth or decay: "Political rivalries can sometimes lead neighboring farmers to a kind of urban competition" (Monbeig 1984: 356). Obviously, no one person can dictate the path of capital through space, but those who possess capital are privileged. In a "new country" with almost no history, capital is almost directly connected to power, and democratic and popular power is not a given: "[the] political steering remains in the hands of a minority who frequently mistakes their own enrichment with that of the collective" (Monbeig 1984, p. 356).

From this notion, Monbeig derives a lesson of the method to his geohistory. Following the lives of certain individuals means to follow the path of the capital that is connected to them: "Their lives can be seen as the history of the pioneer movement" (Monbeig 1984: 355). The urban success of cities also depends on initiative and publicity: "A city is launched as fashion would be, with great bouts of marketing" (p. 357). The occupation of spaces cannot be explained, therefore, only through systemic movements if one seeks more detailed explanations: "key position, the founders' personal action, and collective enthusiasm are factors that explains this fact" (Monbeig 1984: 357):

> The cities in each of the spurs that became small regional capitals [...] did so not only because of more favorable natural conditions but also because of economic and social circumstances: *urban successes were at the same time the triumph of individuals [...].* (Monbeig 1984[1952]: 357–358) [italics added]

The "triumph of individuals" (Monbeig 1984 [1952]: 358) is the triumph of capitalists.

9.10 Conclusions

What did Brazil offer to Monbeig was a question asked by Gilles Lapouge, that gave him an opportunity to evaluate his own experience. He also claimed that he made an association between time and space.

> I discovered space, which is essential for a geographer's other dimensions. I became aware that France is a country of minuscule distances. That is, a change of scale in space, but I also believe that there was a *change of scale in time*, especially for the human geography that I practice. (Monbeig interview with Giles Lapouge 1984, without page) [italics added]

Based on the difference in the consideration of individual actions, some authors have argued that Pierre Monbeig and Fernand Braudel took opposite directions:

> Fernand Braudel, very close to Monbeig, actually went the other way. He sought to substantiate the history of the Mediterranean in permanencies, in very long movements, almost in immobility, as he wrote. The story would somehow subscribe to the long time. [...] Without, of course, ignoring the natural constraints, the limits that are discovered there, and finally the nature of the material upon which men worked, Pierre Monbeig easily demonstrated that geography was a construction of human activity, and therefore of history. (Roncayolo 2006: 126)

I am not sure if the author's position concerning the human capacity to transform space are in opposition. Geohistory is a field of research very suitable for understanding settlement processes, as they appeared in Brazil in the twentieth century or in the Mediterranean in the sixteenth century (colonization of flat land). Braudel describes a situation, in which spaces are the temporary winners in man's struggle to dominate nature. However, I believe that he may have conceived of an increased degree of freedom for man after the Industrial Revolution, related to geographical determinism because he even notes the increased capacity of man in the Mediterranean to overcome natural obstacles (Braudel 1983[1969]).

Monbeig elaborated a unique thinking that was pertinent to Brazil's spatial reality. This reality helped to reshape the legacy of Vidal de la Blache as did Fernand Braudel, both adopting the epistemological perspectives of the space to which they were attached. Pierre Monbeig inserted the Brazilian geographical process into the cartography of the human sciences' advancements in the epistemological and historiographical domain by conceiving a systemic and expansive geohistory, which was also active from the point of view of the development of late and continental capitalism.[20]

Archives Archives Diplomatiques, La Courneve, Paris, France, boxes 443, 444.CAPH, Centro de Apoio à Pesquisa em História (Center for Research in History), University of São Paulo.FFLCH, Booklet commemorating the Pierre Monbeig Chair.Vargas G (1938) Discourse for the inauguration of the New State. Public Domain.

References

Andrews HF (1986) The early life of Paul Vidal de la Blache and the makings of modern geography. Trans Inst Br Geogr 11(2):74–182

Arrighi G (2007) Adam Smith in Beijing. Lineages of the twenty-first century. Verson, London, New York

Azevedo A (1976) Geografia francesa e a geração dos anos setenta. Boletim Paulista de Geografia 50:7–28

Bord JP, Cattedra R, Creagh R, Moissec J-M, Roques G (2009) Elisée Reclus, Paul Vidal de la Blache, la géographie, la cité et le monde, hier et aujourd'hui. L'Harmattan, Paris

Braudel F (1951) La Géographie face aux sciences humaines. Débats et combats Annales de Histoire 6 (4):485–492

Braudel F (1983[1969]) O Mediterrâneo e o Mundo Mediterrânico na época de Felipe II. Martins Fontes, São Paulo

Braudel F (1992) A longa duração. In: Braudel F (ed) Escritos sobre a história. Perspectiva, São Paulo, pp 41–78

Burke P (1997) A Escola dos Annales 1929–1989: a revolução francesa da historiografia. Fundação Editora Unesp, São Paulo

Cassiano R (1970) A Marcha para o Oeste. José Olympio, Rio de Janeiro

Clout H (2003) In the shadow of Vidal de la Blache. Letters to Albert Demangeon and the social dynamics of French geography in the early twentieth century. J Hist Geogr 29(3):336–355

Clout H (2013) Pierre Monbeig. In: Lorimer H, Whiters, CWJ (eds) Geographers Biobibliographical Studies 32, Mansell, London, New York, pp 54–78

Courtot R (2007) Les paysages et les hommes des Alpes du sud dans les carnets de Paul Vidal de la Blache. Méditerranée 109:9–15

Coutinho G (2017) Mário de Andrade e as ideologias geográficas: entidade e território em Macunaíma – o herói sem nenhum caráter. Dissertation (Master in Human Geography) – Faculdade de Filosofia, Letras e Ciências Humanas. University of Sao Paulo, São Paulo

Daix P (1999) Fernand Braudel: uma biografia. Hucitec, Rio de Janeiro, São Paulo

Dantas A (2005) Pierre Monbeig: um marco da geografia brasileira. Sulina, Porto Alegre

Dosse F (1992[1987]) A História em Migalhas: dos "Annales" à Nova História. Editora da Unicamp/Ensaio, Campinas

Dosse F (2004[1998]) História e Ciências Sociais. Edusc, Bauru

Gemelli G (1995) Fernand Braudel. Odile Jacob, Paris

Geraldino CF (2016) A questão da geografia na "Origem das Espécies" de Charles Darwin. Tese (Doctoral thesis). Institute of Geosciences, University of Campinas, Campinas

Gregory D (2000) Cultures of travel and spatial formation of knowledge. Erdkunde 54(4):297–319

Haesbaert R, Pereira SN, Ribeiro G (2012) Vidal, vidais: textos em geografia humana, regional e política. Bretrand Brasil, Rio de Janeiro

Hobsbawm E (1996) The age of revolution, 1789–1848. Vintage Books, New York

Iumatti P, Seabra M, Heidemann D (2008) Caio Prado Jr. e a Associação dos Geógrafos Brasileiros. Editora da Universidade de São Paulo, São Paulo

James PE (1950) Latin America. Cassell and Company, London

Lacoste Y (1989) Lire Braudel. Papirus, Lisboa

Leff NH (1991) Subdesenvolvimento e desenvolvimento no Brasil. Expressão e cultura, Rio de Janeiro

Lévi-Strauss C (1994) Saudades do Brasil. Companhia das Letras, São Paulo

Lima LC (2009) Fernand Braudel e o Brasil. Vivência e Brasilianismo (1935–1945). Editora da Universidade de São Paulo, São Paulo

Limongi F (1988) Educadores e empresarios culturais na construção da USP. Master thesis. Universidade Estadual de Campinas, Campinas

Lira LA (2008) Fernand Braudel e Vidal de La Blache: Geohistória e História da Geografia: Confins [Online]. http://journals.openedition.org/confins/2592, https://doi.org/10.4000/confins.2592

Lira LA (2013) O Mediterrâneo de Vidal de la Blache: o primeiro esboço do método geográfico (1872–1918). Alameda, São Paulo

Lira LA (in press) Pierre Monbeig e a formação da geografia no Brasil (1925–1957): uma geo-história dos saberes. Alameda, São Paulo

[20]One of the main criticism suffered by Braudel in the historiography is the lack of action of his main actors on the scene.

Lira LA, Clerc P (2018) Le Brésil d'Émile Levasseur (1889): Plaidoyer pour un nouveau venu dans l'économie-monde? In: Eden VM et al (eds) Dialogues France-Brésil. Circulations, representations, imaginaires. PUPPA, Pau, pp 69–82
Livingstone DN (2003) Putting science in its place. Geographies of scientific knowledge. University of Chicago Press, Chicago
Loi D (1982) Sur les concepts de description, d'explication et de causalité dans la géographie classique française. Archives from EHGO, Paris, p 1
Loi D (1985) Une étude de la causalité dans la géographie classique française: L'exemple des premières thèses régionales. Espace géographique 14(2):121–125
Loyer E (2015) Lévi-Strauss. Flammarion, Paris
Martins MT (2017) História do pensamento geográfico: formação territorial do Brasil à luz dos projetos territoriais do Exército (1889–1930). Doctoral thesis. Federal University of Uberlândia, Uberlândia
Massi FP (1981) Estrangeiros no Brasil: a Missão Francesa na Universidade de São Paulo. Master thesis. Institute of Philosophy and Human Sciences, State University of Campinas, Campinas
Monbeig P (1952) Pionniers et Planteurs de Sao Paulo. Armand Colin, Paris
Monbeig P (1957[1941]) O Estudo Geográfico das cidades. In: Monbeig P (ed) Novos estudos de Geografia Humana Brasileira. DIFEL, São Paulo, pp 33–77
Monbeig P (1984) Viajem à memória da USP. In: Lapouge G (ed) Entrevista de Pierre Monbeig a Gilles Lapouge. Jornal da tarde, São Paulo, 11 de fevereiro de 1984
Monbeig P (1985[1953]) O Brasil. DIFEL, São Paulo
Monbeig P ([1952]1984) Pioneiros e fazendeiros de São Paulo. Hucitec Polis, São Paulo
Mota LD (1981) A história vivida. O Estado de São Paulo, São Paulo
Müller NL (1951) Sítios e sitiantes no Estado de São Paulo. Boletim da FFCL 132(6):1–215
Neto L (2013) Getúlio: do governo provisório à ditadura do Estado Novo (1930–1945). Companhia das Letras, São Paulo
Peixoto FA (2001) Franceses e Norte-Americanos nas Ciências Sociais Brasileiras 1930-1960. In: Miceli S (ed) História das Ciências Sociais no Brasil, vol 1. Sumaré, São Paulo, pp 477–531
Peixoto FA (2008) Visões de São Paulo: as cidades de Lévi-Strauss. Bastide e Monbeig, Revista da Biblioteca Mário de Andrade 64:82–106
Pericás LB (2016) Caio Prado Jr.: uma biografia política. Boitempo, São Paulo
Petitjean P (1991) Dimensão cultural, influências ideológicas e imagens científicas Franco-Brasileiras, 1850/1940. In: Parvaux S, Revel-Mouroz J (eds) Images reciproques du Brésil et de la France. Actes du Colloque organisé dans le cadre di projet France-Brèsil, IHEAL, Paris, pp 845–860
Petitjean P (1996a) As Missões Universitárias Francesas na Criação da Universidade de São Paulo (1934–1940). In: Hamburguer AI et al (eds) A Ciência nas relações Brasil-França (1850–1950). Edusp, Fapesp, São Paulo, pp 259–330
Petitjean P (1996b) Entre Ciência e Diplomacia: a organização da influência científica francesa na América Latina, 1900–1940. In: Hamburguer AI et al (eds) A Ciência nas relações Brasil-França (1850–1950). Edusp, Fapesp, São Paulo, pp 89–120
Petitjean P (1998) Autour de la mission française pour la création de l'Université de São Paulo (1934). In: Petitjean P, Jami C; Moulin AM (eds) Science and empires: historical studies about scientific development and European expansion. Kluwer Academic Publishers, Dordrecht, London, Boston, pp 339–362
Pirenne H (1963[1933]) História econômica e social da Idade Média. Martins Fontes, São Paulo
Reis JC (2000) A Escola dos Annales e a Inovação em História. Paz e Terra, Rio de Janeiro
Robic MC (2000) Le Tableau de Géographie de la France de Paul Vidal de la Blache. CTHS, Paris
Robic MC (2004) Un système multi-scalaire, ses espaces de référence et ses mondes. L'Atlas Vidal-Lablache. Cybergeo 265. 10.4000/cybergeo.3670
Romano R (1997[1995]) Braudel y nosotros: reflexiones sobre la cultura histórica de nuestro tiempo. Fondo de Cultura Económica, Ciudad do México
Roncayolo M (2006) Tempos geográficos e construção dos espaços na análise de `Pierre Monbeig. In: Salgueiro HA (ed) Pierre Monbeig e a Geografia Humana Brasileira: a dinâmica da transformação. Edusc, Bauru, pp 117–128
Rudwick MJ (2014) Earth's deep history: how it was discovered and why it matters. University of Chicago Press, Chicago
Salgueiro HA (2006a) Pierre Monbeig e a Geografia Humana Brasileira: a dinâmica da transformação. Edusc, Bauru, São Paulo
Salgueiro HA (2006b) Introdução à Primeira Parte. In: Salgueiro HA (ed) Pierre Monbeig e a Geografia Humana Brasileira: a dinâmica da transformação. Edusc, Bauru, pp 17–35
Salgueiro HA (2006c) Introdução à Segunda Parte. In: Salgueiro HA (ed) Pierre Monbeig e a Geografia Humana Brasileira: a dinâmica da transformação. Edusc, Bauru, pp 89–115
Schwartzman S (1979) Formação da Comunidade Científica no Brasil. Ed. Nacional, São Paulo, Financiadora de Estudos e Projetos, Rio de Janeiro
Sousa Neto M (2012) Planos para o Império: os planos de viação do Segundo Reinado (1869–1889). Alameda, São Paulo
Suppo HR (2000) A Política Cultural da França no Brasil entre 1920 e 1940: o direito e o avesso das missões universitárias. Revista de História 142–143:309–345
Suppo HR (2002) La politique culturelle française au Brésil entre les années 1920–1950. Doctoral thesis. Université Paris III—Sorbonne Nouvelle, Paris
Théry H, Droulers M (1991) Pierre Monbeig: un géographe pionnier. IHEAL, Paris

Travassos M (1938) Projeção continental do, Brasil edn. Nacional, Brasil
Vidal de la Blache P (1896) Le principe de la géographie générale. Annales de Géographie 5(20):129–142
Vidal de la Blache P (1911) Les genres de vie en géographie humaine. Annales de Géographie 112 (20):289–304
Vidal de la Blache P (1914) Sur l'esprit géographique. Revue Politique et Littéraire (Revue Bleu) 52 (18):556–560
Vidal de la Blache P (1984[1981]) Atlas général Vidal-Lablache: histoire et géographie. Armand Colin, Paris
Whithers CWJ (2007) Placing the enlightenment. University of Chicago Press, Chicago

10 How International was the International Geographical Congress in Rio de Janeiro 1956? On Location and Language Politics

Mariana Lamego

Abstract

The chapter focuses on scientific communication by addressing the question: how international was the eighteenth International Geographical Congress (IGC) held in Rio de Janeiro in August 1956. The question is addressed, firstly, considering the location politics of the unique case of an international geographical congress of the International Geographical Union (IGU) held in South America, focusing on some practical, theoretical, and political impacts. Secondly, the focus is on language aspects, once the Rio IGC can be seen as a turning point in language diversity in IGU toward the consolidation of English as the main, and almost exclusive, language of international scientific communication. Considering the present challenge of critically discussing the crescent hegemony of English as the only possible language of scientific communication, the chapter concludes stressing how the study of some experiences of multilingualism in past can help us in proposing other possibilities of scientific communication in a more symmetric power position and condition.

Résumé

Le chapitre se concentre sur la communication scientifique en abordant la question: quelle a été l'internationalité du 18e Congrès International de Géographie (CIG) qui s'est tenu à Rio de Janeiro en août 1956. La question est d'abord abordée en tenant compte de la politique de localisation du cas unique d'un congrès géographique international de L'Union Géographique Internationale (UGI) s'est tenue dans une Amérique du Sud, en se concentrant sur certains impacts pratiques, théoriques et politiques. Deuxièmement, l'accent est mis sur les aspects linguistiques, étant donné que le CIG de Rio peut être considéré comme un tournant dans la diversité linguistique de l'UGI vers la consolidation de l'anglais en tant que langue principale et presque exclusive de la communication scientifique internationale. Considérant le défi actuel de discuter manière critique l'hégémonie du croissant de l'anglais comme seule langue possible de communication scientifique, le chapitre conclut en soulignant comment l'étude de certaines expériences du multilinguisme dans le passé peut nous aider à proposer d'autres possibilités de communication scientifique dans un position et état de puissance plus symétriques.

M. Lamego (✉)
Universidade do Estado do Rio de Janeiro (UERJ), Rio de Janeiro, Brazil
e-mail: marilamego@gmail.com

B. Schelhaas et al. (eds.), *Decolonising and Internationalising Geography*, Historical Geography and Geosciences, https://doi.org/10.1007/978-3-030-49516-9_10

Resumen

El capítulo se enfoca en la comunicación científica al abordar la pregunta: ¿Cuán internacional fue el 18° Congreso Geográfico Internacional (CGI) celebrado en Río de Janeiro en agosto de 1956. La pregunta se aborda en primer lugar considerando la política de ubicación del caso único de un congreso geográfico internacional de la Unión Geográfica Internacional celebrada en América del Sur, centrándose en algunos impactos prácticos, teóricos y políticos. En segundo lugar la atención se centra en los aspectos del lenguaje, teniendo en cuenta que el Río CGI puede ser visto como un punto de inflexión en la diversidad lingüística en la UGI hacia la consolidación de Inglés como el principal, y casi exclusivo, lengua de comunicación científica internacional. Considerando el desafío actual de discutir críticamente la creciente hegemonía del inglés como el único idioma posible de comunicación científica, el capítulo concluye enfatizando cómo el estudio de experiencias de multilingüismo en el pasado puede ayudarnos a proponer otras posibilidades de comunicación científica en un posición y condición de potencia más simétrica.

Resumo

A comunicação científica é o foco deste capítulo no qual é colocada a questão do quão internacional teria sido o 18o Congresso Internacional de Geografia (CIG), realizado na cidade do Rio de Janeiro em agosto de 1956. A questão é abordada primeiramente considerando a política de localização do caso único de um congresso geográfico internacional da União Geográfica Internacional (UGI) realizado na América do Sul, com foco em alguns impactos práticos, teóricos e políticos. Em segundo lugar, a ênfase é dada aos aspectos linguísticos do congresso, uma vez que o CIG do Rio pode ser visto como um ponto de virada na diversidade de idiomas da UGI em direção à consolidação do inglês como principal e quase exclusivo idioma da comunicação científica internacional. Considerando o atual desafio de discutir críticamente a crescente hegemonia da língua inglesa na comunicação científica, o capítulo conclui destacando como o estudo de experiências de multilingualismo no passado pode nos auxiliar na proposta de outras possibilidades de comunicação científica em posições e condições de poder mais simétricas.

Keywords

International Geographical Congress · Rio de Janeiro · International Geographical Union · Scientific communication · Location politics · Language diversity

10.1 Introduction

Scientific meetings constitute a very important part of our academic life. It is a place where we can exchange our disciplinary view, get in contact with new ideas, and meet new and old colleagues strengthening our academic ties. In an increasingly way, scientific meetings, whether in the form of conferences, congresses, colloquia, or symposia, are studied as crucial elements in the shaping of the disciplinary knowledge by diverse approaches (Withers et al. 2006, 2008; Withers 2010a, b; Robic 2010). These approaches portray scientific meetings as nodes in complex networks of bodies and artifacts, engaged by academic communities with different interests and institutional spaces. They are not only key-sites of knowledge production (Craggs and Mahony 2014); they represent entirely Livingstone's model of scientific knowledge circuit of production, circulation, and reception (Livingstone 2007, 2010), constituting a convergence space, as described by Routledge (2000).

Among diverse sort of encounters, some will adopt the international label, which supposedly means that their range is beyond the nation's limits and open to other nations, constituting a place of interaction as well as a contention field between heterogeneous academic cultures from the globe. Thanks to theoretical approaches (Shimazu 2012, 2014; Craggs 2014), encounters labeled international can also be taken "as geopolitical events through which distinctive political positions and identities were staged and performed" (Craggs 2014: 39).

Thus, once considering the international meetings as a scientific knowledge circuit, a convergence space, a dispute field as well as a geopolitical event, one cannot pretend that the language employment or the choice of official language (or languages) is just a necessary or even a natural procedure in order to guarantee a satisfactory communication within a multilingual international scientific community. Speaking as a Latin American geographer, I understand in my practice that language is not merely a matter of communication; language is a matter of positionality and consequently a matter of representation in a supposedly international community.

In this paper, I focus on this crucial aspect of scientific communication by addressing the question: how international was the eighteenth International Geographical Congress (IGC) held in Rio de Janeiro in August 1956? The eighteenth IGC in Rio was the only International Geographical Congress in the whole history of IGU taking place in South America,[1] and it was the last time six official languages were accepted. Therefore, my ultimate aim is to give some attention to this crucial point for our international geography community, emphasizing the communion of geographical perspectives in more symmetric power relations.

The issue was in the focus again in recent discussions on internationalism and its rhetorical application in all sort of meetings (Legg 2014; Hodder et al. 2015; Hodder 2015). These debates refer to actual discussions about internationalization and monolingualism in scientific communication via systematic adoption of English as the *lingua franca*.

The question about the internationality of the IGC will be addressed following two aspects. Firstly, I present aspects of the Rio meeting with regard to location politics at stake in the organization of the first and the only IGC in Latin America, highlighting some of the practical, theoretical and political impacts. In the second path, my focus is on language aspects, considering that the Rio IGC can be seen as a turning point in language diversity in IGU toward the consolidation of English as the main, and almost exclusive, language of international scientific communication in IGC.

10.2 Why Rio? Location Politics on IGC Venue Choice

> A realização do Congresso Internacional de Geografia no Rio de Janeiro havia sido lembrada para anos atrás, mas o destino reservou-nos exatamente essa oportunidade feliz, quando se processa com energético entusiasmo uma transformação de infra-estrutura econômica, oferecendo campo largo de investigações. Os geógrafos brasileiros (…) rejubilam-se ante a oportunidade de receber e de conviver no Brasil ao lado de companheiros vindos de todas as partes da terra, de todos os continentes, pertencentes a todas as raças e que aqui se irmana e confraternizam no trabalho e nos ideais que só a ciência inspira.[2] (opening speech of Jurandir Pyres Ferreira, head of the local organizing commission of the 18th IGC, August, 1956—in: Comptes Rendus 1959: 136–137)

> Esta assembleia de mestres, aceitando o convite do Brasil para realizar seus conclaves na orla da baía de Guanabara, num dos mais belos cenários naturais do mundo, não se deixou conduzir apenas por aquela motivação estética, que faz do geógrafo um contemplativo, para quem a beleza panorâmica é

[1]IGU promotes International Geographical Congress each four years, and also promotes regional conferences and comission meetings at different frequencies. Considering IGU regional conferences, they were held in South American countries such as Havana 1995, Buenos Aires 2005, Santiago de Chile 2011, and Thematic Conference La Paz 2017.

[2]The realization of the International Geographical Congress at Rio de Janeiro had been remembered for years, but the destiny has reserved us exactly this happy opportunity, when a huge transformation in country infrastructure offers a broad investigation field. (…) Brazilian geographers (…) rejoice for the opportunity to receive and live in Brazil alongside with comrades coming from all parts of the world, from all continents, belonging to all races, which here unites them and fraternizes in the work and in ideals that only science inspires.

> um estado de poesia: atendeu igualmente à circunstância de que a ciência geográfica no Brasil contemporâneo atingiu a sua maturidade. (…) Enquanto o mundo se entrelaça, na composição dos interesses recíprocos, cada país necessita ser conhecido e conhecer os outros, para a perfeita ajustagem no concerto das nações modernas. Eis a alta finalidade deste congresso. (…).[3] (opening speech of Brazilian President Juscelino Kubitschek de Oliveira, August 1956—in: Comptes Rendus 1959: 154–155)

The study of international meetings as geopolitical events enables a possible new understanding of the role played by congress locations (Shimazu 2012). The question, *where* a meeting takes place, deeply influences individual and institutional participation, actor's performances, media narratives, as well as the spoken languages too. Furthermore, the meeting location —including its choice—is embedded into geopolitical aspects, revealing social and political constructions of the "international" label the congress is displaying.

Regarding the selection criterion of the IGC venue, according to Pinchemel (1972), the Executive Committee of IGU—whose members should meet at least once during congress breaks—must consider numerous aspects, from political and economic stability of the host country, costs of transportation and hotel facilities to the situation of scientific development and the presence of enough geographers to assure the preparation and development of the congress (Pinchemel 1972: 30).

Considering these two last requirements specification on scientific development and the recognition of a substantive body of geographers at the host country, the extract of the opening speeches by the geographer Jurandir Pyres Ferreira, President of the National Council of Geography, and the Brazilian newly elected President, Juscelino Kubitschek, reveals that at that time Brazil lived a real enthusiastic moment in political and economic development. Juscelino Kubitschek has taken office as President in January 1956 for a four-year term, elected with a government program that announced broad political and economic reforms for the country. In fact, there was a continuity of the policy for the development of the industrial sector of the country already started under the previous mandate of Getúlio Vargas, but the so-called Plan of Goals of JK government deepened the reforms to the transport and energy sectors. The JK government also marks the transfer of the country's capital, previously in Rio de Janeiro city to Brasilia, a city that was to be completely built in the central geographical region of the country. The enthusiasm for the new policies ahead for the next four years in the country was not only felt by the productive sectors of private enterprise. In scientific realm, the period expressed the prominence of geography in the country and the recognition of the strategic role played by geographical knowledge to the government auspices. Some institutions such as the Brazilian Institute of Geography and Statistics (IBGE), created during Getúlio Vargas period in 1934, were empowered from 1956 onward. Created in 1937 within the IBGE's administrative structure, the Geography National Council (CNG) has established itself as a specialized body to effect thorough and systematic knowledge of the country, coordinating various initiatives, such as field trips, territorial surveys, and regionalization, among others (Aranha 2014: 2). The Brazilian Research Center on Geography[4] (CPGB), established in 1951 at the National University of

[3]This council of masters, accepting the invitation of Brazil to hold its conclaves on the shores of Guanabara Bay, in one of the most beautiful natural settings in the world, was not only guided by that aesthetic motivation, which makes the geographer a contemplative, for whom a panoramic beauty is a state of poetry, it also attended to the circumstance that geographical science in contemporary Brazil reached its maturity. (…) As the world intertwines, in the composition of reciprocal interests, each country needs to be known and to know others for the perfect adjustment in the concert of the modern nations. This is the high purpose of this congress.

[4]The Brazilian Research Center on Geography was created thanks to the effort of Hilgard O'Reilly Sternberg, a key figure inside the historiography of Brazilian geography. At IGU, he was the Vice President of the Executive Committee from 1952 to 1956 and became the Executive Secretary of the Local Organizing Committee. As a true recognition of success of the Rio IGC, he became the first Vice President of IGU from 1956 to 1960.

Brazil in Rio de Janeiro, was the largest geographer training and research center in the country at that time. Military institutions also played an important role in the development of Brazilian territorial knowledge such as the National War College, established in 1949 as an Institute for Advanced Studies in Politics, Defense, and Strategy, as a part of the Brazilian Ministry of Defense.

Thereby, at the mid-twentieth century, although Brazilian geography was still relatively young in terms of academic institutionalization, the country already had a great body of researchers in geography as well as related fields which could satisfy the requirements specification. The above-mentioned institutions—IBGE, CNG, CPGB, and ESG—not by chance are among the main organizing institutions of the eighteenth IGC in 1956.

It was at the eighth General Assembly held in Washington DC during the seventeenth IGC in 1952 that Rio de Janeiro was presented as candidate to hold the next meeting in 1956. According to Pinchemel (1972: 29), the city of Rio de Janeiro would have competed "friendly" with the cities of Edinburgh and Vienna. Nevertheless, the choice of Rio de Janeiro as the host of the eighteenth IGC should not be understood only in terms of the alleged well development of our geographical knowledge production or even considering the elated phase on Brazilian political and economic affairs. It must also be considered based on the IGU internal politics of congress location put into practice after the Second World War, which in its turn reveals a strategy of "déseuropéanisation", in Pinchemel terms (1972: 29), regarding not only the meeting venue considering Washington (1952) and Rio de Janeiro (1956) but also the admission of new members outside the European domain. It is worth pointing out that the IGU internal politics of congress location is connected in a broader phenomenal scale, which considers political, economic, social, and cultural aspects varying from congress to congress.

The notable shift in location politics of IGC is stressed by some authors interested in the relations between the spatiality of IGU's international congresses and the geopolitics involved in the choice of country hosts and new members association (Pinchemel 1972; Volle 1996a, b; Robic 2010). Commonly, those researchers point to the fact that until 1952 the meetings were concentered in the European territory. It was in the seventeenth IGC in Washington that we find a change toward a more international perspective, mainly by an opening admission directed to member countries from Asia and Latin America, just after a long, turbulent, and contentious period of political conflicts embedded in scientific proposals of the IGU in the interwar period (Rössler 1990).

As pointed by Robic (2010: 39), there was an "extraordinaire transformation de la carte de pays membres de l'UGI qui s'est produite, par saccades, après la Second Guerre mondiale. Au nombre de 30 avant-guerre, l'effectif a doublé de 1949 à 1964 (de 31 à 62), enregistrant le volontarisme de présidents de l'Union, la réhabilitation des pays vaincus (avec les tensions de la Guerre froide) et surtout, le démantèlement des empires coloniaux".

Those aspects related above give us the contextualization we need to raise possible reasons for the Rio's choice, but it is important to think about its practical consequences as well, such as the influence upon the balance in communications among participants from the host country and from other nations, or the election of the themes and proposal questions, which guide the scientific communications, as well as the symposia within the IGC (Pumain 1972). Considering my aim, here I want to focus on two practical consequences of the politics of location: firstly, the participation dynamics of countries' members, considering that depending on where the meeting will take place, new actors can appear as well as others can be in absence; and secondly, the tendencies on language use, which will be developed in the next two sections.

10.3 Who is Coming Anyway? Congress Organization and Participant Dynamics

Following the choice of Rio de Janeiro as the venue for the eighteenth IGC, the first step was the institutionalization of a local organizing committee charged by the total organization of the meeting, which, according to IGU general rules, should submit its plans previously for approval to the Executive Committee at least two years before the meeting (Dubois 1972). Sponsored by the Brazilian Geography National Council, the IGU National Committee published in 1954 the document which allowed the creation of the local committee with its attributions and structure. Constituted by three boards—deliberative, advisory, and executive—the local committee was mainly composed by geographers and professionals from the staff of the Brazilian Institute of Geography and Statistics. Within the sub-committees, one can find actors from other institutions involved in the organization, such as the military school, army, and governmental ministry. The composition of the local committee with the presence of high-ranking military together with civilian geographers helps us to understand the weight of the meeting, not only for the geographical community, but also to the national political sphere as well.

The IGC venue in the city was a place directly connected to the local committee composition and its military feature. The main meeting venue was the Naval School, localized at Villegagnon Island, in the Guanabara Bay. Thanks to the shift of the school holidays from July to August, the whole structure of the Naval School was completely available for the congress. The Naval School became the meeting headquarters, as pointed out by the US geographer Charles B. Hitchcock (1957) in his account on the congress one year later in the *Geographical Review*. The facilities were used for all paper sessions and commission meetings. The student lodging was used by male members with no charge, and the student restaurant was used by participants during the congress. There were also room for the press, a post office, exchange agency, newspaper stand, and a souvenir shop. This mini city appearance of the eighteenth IGC venue is mentioned in the congress proceedings as a strategy to overcome the inconvenient insulation, yet the text also highlights the advantages of being on an island such as “un rapprochement plus étroit entre les participants” (Comptes Rendus, Comité National du Brésil 1959).

Looking at the delegates’ geographical distribution in the eighteenth Congress, one can see the increase in numbers from some countries compared to previous congresses. It was the case of nations from Africa, Asia, and South America that for the first time could send members to attend an IGU international meeting, such as Angola, Ethiopia, and Guatemala or even for the second time such as Algeria, Bolivia, Haiti, Iceland, Indonesia, Iraq, Pakistan, Panama, and Puerto Rico (Kish 1972). The participation of these countries reinforces the argument that IGU Executive Committee was engaged in a politics of enlarging its world map. In fact, it was during the congress in Rio de Janeiro at the ninth IGU General Assembly that the delegates approved the creation of a new member category. This new category, called “associate members”, allowed the participation of countries that would not have the basic requirements to become full IGU members, usually connected with a National Geography Committee (Dubois 1972). According to the text published at the proceedings,

> Countries with a small number of geographers may seek election by the General Assembly as Associate Members. Associate Members shall have no vote on administrative matters in the General Assembly: their financial obligations shall be decided by the Executive Committee [provisionally fixed at $100.00 (U.S.) a year]. Applications for Associate Membership shall be signed by not less than three geographers resident in the country concerned, one of whom shall be named as their representative. The list of Associate Members shall be reviewed at each meeting of the General Assembly. (Comptes Rendus, Comité National du Brésil 1959)

The introduction of this new category of members allowed a possible participation of poor countries—resulting from the political emancipations that took place after the Second World War—to the new map of the IGU members. That was the case for six countries, which applied for associate membership: Iraq, Kenya, Nigeria, Sudan, Uganda, and French West Africa—which disintegrated itself after two years in 1958.

The period of a strong orientation toward an internationalization of the IGU was clearly visible at the Rio IGC, but we have to notice: it was not only a matter of participation, but also a matter of representation. The case of South American delegates illustrates this shift, because it is possible to see a substantial increment in its representation. At the seventeenth Congress 1952 in Washington, South America had 6% of the total number of participants, and in 1956, this number increased to 67%, as pointed out by Hitchcock (1957: 118–119). Surely, the decrease in distance, and consequently a decrease in travel costs can explain this shift, but another kind of explanation, not so linked with geographical distance but actually connected to geopolitical issues, can explain the arrival of new actors at the eighteenth IGC in Rio.

The case of USSR participants in Rio de Janeiro shows clearly how geopolitical issues obviously influenced international congresses. In 1956, the Soviet delegation was the sixth in numbers, composed by 12 geographers from the USSR Science Academy and from the University of Moscow. Absent from the three previous congresses, the delegation landed in Rio with a very important mission: to present its application for regular membership in the ninth General Assembly. The mission was well accomplished, out of possible 26 votes, by the heads of the present delegations, the application received 23 votes. The USSR delegation was not the only one applying for regular membership from the socialist world, two more applications from Hungary and the German Democratic Republic were on the agenda.

The application of the Soviet delegation did not pass unnoticed by the main newspaper at the time, *O Globo*, which dedicated a half page report on the subject. Under the suggestive title "In Geography, the United States is with Russia", published on 12th August, the report comments the return of the Soviets to the international congress of geography after 30 years of absence. The newspaper article narrates exactly the results of the General Assembly with the voting of the 38 member countries. The number differs from the 26 published in the minutes of the Assembly, but 26 is the number of delegates presented at the meeting, the other 12 votes were made by letter by the absent members. The report brings the result of the vote, pointing out that the Soviet Union received three votes against its admission, but none of them had been from the US delegation. Wallace Atwood, the head of the US delegation, was interviewed and, according to the report, had demonstrated immense satisfaction in having the Soviets back to the IGU, highlighting that the country should have been admitted 30 years before, and finally stating that "the geographical science does not acknowledge political frontiers" (O Globo, August 10th 1956).

The report also brings information on the application of the German Democratic Republic, pointing that it was accepted by unanimity. The report presents a fatal error. Actually, the East German delegation did not get the application approved and received just 15 votes of 26 present in total. As a consequence of the confusing result, IGU President Dudley Stamp proclaimed that the "election of the German Democratic Republic [shall] be regarded as *sub judice*" having the final decision be postponed to the next meeting of the General Assembly, which was held in Stockholm in 1960. The admission of the GDR as a regular member of the IGU was a point of contention among its members. In contrast to Atwood's statement, political frontiers and geopolitical affairs did exist in the minds of the IGU members, especially considering that since 1952 the Federal Republic of Germany was already officially belonging to IGU, appearing in the bulletin under the name of Germany, without any mention of the division of the country (Volle 1996a).

The media covering is a very illustrative tool in order to visualize how some narratives and performances were translated highlighting the international rhetoric. At the same time, it is also possible to understand how international scientific meetings were potentially used to reinforce political questions underlying scientific practices. As pointed out by Craggs (2014) and Shimazu (2012), conference performance is an object of media translation and tells more about geopolitics than about the meeting itself, and that is why the press is also an important vehicle of some narratives that circulated and helped us to interpret and create some ideas about the congress. Moreover, according to Dalby (1996), media presence at international conferences does not solely translate and transmit political agendas, but also often sets them by highlighting the views of certain delegates over others.

If location politics is a very important component in understanding the international rhetoric in IGC and IGU practices, language politics is the second important one.

10.4 What Languages Will Be Spoken? International Geography and Language Politics

In 2000, human geographer Claudio Minca wrote a short but potent paper stating that "the boundaries as well as the rules/coordinates of what passes for 'international' debate within our discipline are determined from within the Anglo-American universe" (Minca 2000: 287). Taking as example precisely a geography conference held in Venice in 1999, understanding it as a geographical praxis embedded in its power relations, the author criticizes the domination of the English language in plenaries and communications although the meeting was attended by more non-native English speakers than native ones.

Minca's paper prompted a long-term and passionate discussion in which many geographers from different countries positioned themselves (Samers and Sidaway 2000; Gutierrez and Lopez-Nieva 2001; Desforges and Jones 2001; García-Ramon 2003; Aalbers 2004; Kitchin 2005; Aalbers and Rossi 2009). Although very diverse and sometimes featuring antipode positions, the discussants agreed in the fact that the hegemony of the Anglo-American geography, reinforced by the use of English as the "universal" language of scientific communication particularly since Second World War, is responsible for the fading, and even the total exclusion of other geographical and linguistic communities.

Many years before this fierce debate, Whitehand and Edmondson showed already in their 1977 well-known paper a reorientation in geographical communication between the 1950s and 1970s with a decrease in French and German language and an increase in English. The reason for this 'balance of power' was according to the authors the irruption of the quantitative revolution within the discipline. Whitehand and Edmondson's analysis had its focus on transformations inside the discipline that could explain the shift toward Anglo-American language hegemony. Although it is possible to relate the impact of quantitative revolution, it is not possible to give so much weight to an internalistic explanation. It is necessary to go beyond to realize how connected the language issue in a discipline such as geography with changes in new world order really is.

Within this debate, we should not forget the changes of geopolitical balance of world power and the role of some twentieth century intergovernmental organizations, such as the League of Nations, founded in 1920 and its substitute the United Nations, founded in 1945. These organizations helped yielding new meanings of "international" which definitely impacted the necessity of the adoption of a global *lingua franca*. The "international" was not a given category or scale (Hodder 2015: 40), so the movement toward its definition is not out of place, and although it relies on the view from nowhere rhetoric (Shapin 1998), in fact, it is very situated and positioned. Behind the meaning of "international" are different and sometimes competing ideals, which defend what and how, and even where should international be represented, constituted, and moreover, in what language it should speak.

In this way, the so-called international scientific meetings could be the very illustrative examples of how some rhetorical strategies on "the international" take place, and the understanding of the dynamics on language adoption could be an eloquent subject of analysis. Some authors dedicated words and pages on the matter of language just regarding the case of IGU international meetings (Harris 2001; Robic 2010; Volle 1996a, b). Among these scholars, it is easy to identify two clearly different positions regarding the issue on language: On the one hand, there is an obvious focus on monolingualism, as a natural outcome of the need to increase the effectiveness of communication in a so-called international community, which only increases year by year. On the other hand, we can recognize the existence of this trend, but not as a natural outcome, following the principle "the lesser of two evils". On contrary, some authors propose a critical positioning regarding this important matter and advocate for multilingualism in scientific communication.

Chauncy Harris, IGU Secretary General and Treasurer from 1968 to 1976 (Volle 1996b) evaluated the rise and spread of English (Harris 1998, 2001). He argues that after the Second World War, the improvement on mobility allowed a broader worldwide participation prompting the reduction of congress language in order to facilitate intercommunication.

Harris argument is close to the first set of scholars described above and seems to be close to Whitehand and Edmondson (1977) too. In both papers, we can find the naturalization of the monolingualism trend. This idea is supported by a common argument, that the decrease of a language democracy is a necessary side effect, which guarantees the efficiency of intercommunication. Desforges and Jones explain the importance of this argument, asserting that attention must be payed "to the politics of language, or in other words, the way in which the use of language can imbue meaning and power to certain social and spatial practices" (Desforges and Jones 2001: 262).

The other pole of this discussion is mainly dominated by the French community, with regard to IGU history especially by Robic (1996, 2010) and Volle (1996a, b). For both, the language policy of the IGU should be analyzed in close relation to the history of the IGU itself, which in turn, is embedded in its political, economic, social, and cultural context. Volle (1996b) investigates the language adoption from the first IGC in 1871 until 1992. In her analysis, Volle shows what characterizes as 'le passage d'une prodigalité initiale, liée au polylinguisme, à l'uniformité d'une langue dominante' (1996b: 74).

Until 1956, there were at least six official IGU languages: English, French, German, Italian, and Spanish. These official languages were added by Flemish at Anvers in 1871, Polish at Warsaw in 1934, and Portuguese at Lisbon in 1949 and at Rio in 1956. As pointed out by Volle (1996b), the sample of official languages is in fact a demonstration of the European hegemony. The scenario of European hegemony will change after the beginning of the internationalization politics of IGU, explored in the section above. The opening of the IGU to non-European countries, especially in the beginning of the mid-twentieth century, is paradoxically accompanied by the narrowing of linguistic plurality.

Therefore, one important question comes up: how can we explain this paradox, and how was eighteenth IGC in Rio involved? It is important to look at some of the papers, presented in 1956, regarding the used languages and the publication in the congress proceedings. Table 10.1 shows the language employed in presentations and the nationality of the presenters which was made by the information from the proceedings of the eighteenth IGC.

Almost 350 abstracts were submitted, and 267 were presented during the ten days of the congress. In the call for papers, published two years before the meeting, the guidelines told to deliver the abstracts in French or in English, but the paper and the oral presentation were accepted in those six languages already mentioned. This diversity can be seen in the distribution of papers per session and the language delivered. Not surprisingly, the highest percentage is from papers presented in English with almost 39% of the total number, because it was the language used not only by Anglophone presenters, but by

Table 10.1 Language usage by nationality (Source: Comptes rendus du dix-huitième congrès international de géographie, Comité National du Brésil 1959)

Nationality	FRA	ENG	GER	POR	SPA	ITA	TOTAL
Angola				1			1
Argentina	2	1			4		7
Australia		2					2
Austria			2				2
Belgium	2						2
Brazil	1	4	2	38			45
Canada	2	4					6
Chile					1		1
China		2					2
Cuba		2					2
Czechoslovakia		1					1
Denmark		4					4
Egypt	1						1
Ethiopia		2					2
Finland		1					1
France	50						50
Germany		6	14				20
Greece	1						1
Hungary			1				1
India		6					6
Iraq		1					1
Israel		3					3
Italy	3					10	13
Japan	1	8					9
Mexico					2		2
Netherland	1	3					4
Pakistan		4					4
Poland		6					6
Portugal	2			2			4
Romania	2						2
Spain					5		5
Sweden		2					2
Switzerland	1	6					7
UK		11					11
Uruguay					6		6
USA		19					19
USSR	3	5					8
Venezuela					2		2
Yugoslavia	1	1					2
Total	73	104	19	41	20	10	267

delegates from other countries such as Denmark, India, Poland, Egypt, Switzerland, Pakistan, Iraq, Sweden, Czechoslovakia, Yugoslavia, China, Israel, Ethiopia, and Japan. French was the second language (27.5%) used not only by French presenters but also by scholars from Canada, Netherlands, Greece, Egypt, Belgium, Japan, Romania, and Soviet Union. Portuguese was the third most used language with 15.5%. This percentage is naturally due to the massive participation of Brazilian researchers. There was around 7% in German and Spanish, and 4% in Italian, which corresponded to the Italian delegation presentations entirely (Comptes rendus du dix-huitième congrès international de géographie, Comité National du Brésil 1959).

Considering the data displayed at the table, regarding the six most numerous delegates by country whose language was among the six official languages accepted, some inferences can be made. All presentations made by French delegates were presented in French. All presentations made by UK delegates were presented in English. All presentations made by the USA delegates were presented in English. Not all presentations made by Brazilian delegates were presented in Portuguese, there were four in English, two in French, and one in German. Not all presentations made by Italian delegates were presented in Italian, there were two presentations in French. Not all presentations made by Argentinian delegates were presented in Spanish, there were two in French and one in English. The conclusion one can have by analyzing the numbers at the table is that the monolingualism seems to be founded substantially among native speakers of English and French since most of all other nationalities, including those whose mother tongues are listed as official (the case of the Germans, Brazilians, Portuguese, Argentines, and Italians) used multilingualism as a form of expression in scientific communication.

The most euphoric phase (Volle 1996b) in the linguistic democracy of IGU was at the same time accompanied by first complaint on multilingualism. The original argument as an advance in cultural exchange was now qualified as disadvantage, with negative consequences to the motives of an international congress in geography, with rigorous, efficient, and rational order. Chauncy Harris, who focused on the decline of multilingualism within IGU, was interviewed by Rössler (1996). He described the great diversity of languages as "one of the continuing problems of scientific communication among countries" (Harris to Rössler 1996: 301). Talking precisely about Rio's IGC, Harris narrates his unpleasant experience, as co-chairmen of the commission on economic geography, along with Dino Gribaudi from Italy:

> Papers were presented in six languages. If the paper was in Italian, Spanish or Portuguese, Gribaudi presided. If the paper was in German, French or English, I presided. Two successive papers were presented by a Brazilian geographer and a German geographer, the first in Portuguese and the second in German. During the first paper the audience consisted almost solely of Brazilians. When time for the second paper came, this first audience walked out and an entirely different audience of Western Europeans came in. This was not international communication among geographers. It was a series of sub-groups talking to other members of the same linguistic sub-group. (Harris to Rössler 1996: 301)

We can find several counter-arguments to face the eloquent narrative made by Harris, an English native speaker. One argument derives from the various participations of Brazilian geographers with a bilingual or even trilingual background. This was a good side effect of the French influence within early Brazilian geography. The theoretical and methodological affiliation to post-war Anglo-American geography is another development. Therefore, we must seriously consider Harris' arguments as a Gordian knot regarding scientific communication. The almost exclusively sub-group intercommunication unfortunately is not a rare phenomenon at international congress of geography until nowadays.

In contrast to Harris' experience, the actual situation is for instance characterized by an internationally labeled congress in geography, with Chinese researchers speaking to Chinese researches in English (Cameron et al. 2016).

As mentioned before, the pro argument in scientific communication seems to seduce primarily those who already occupy privileged positions and to whom the adoption of two languages, as long as it includes their native language, is presented as the only solution. Indeed, "linguistic hegemony is a form of power that empowers some while disempowering others" (Short et al. 2001: 1).

Backing to the IGU language politics, a true change of mentality seems to operate within the IGU, in which one notion of diversity gives way to another of unity, where the practice of relatively symmetrical exchanges between peers begins to be replaced by the imposition of a linguistic choice. At the IGC 1960 in Stockholm, the panorama turns out completely. Instead of six languages, only two, English and French were used. This was described by Volle (1996b: 77) as a period "où s'instaure une attitude officielle beaucoup plus rigide".

This babel tower nature of the eighteenth IGC got media attention during the meeting press cover. The foreigner guests became news in reports that covered since the arrival of delegation, passing through some visiting in touristic places in the city, till the gala dinner with the presentation of folklore groups. The social media was interested in the reactions of the visitors and, in general, the reports tried to create an integration idea highlighting the importance of Brazil as a stage for this integration, being multilingualistic depicted as not a real problem.

An interesting report, entitled "The purpose is geography but greater is the interest in life", published on 10th August at *O Globo* journal pages, covers the very intense first day of work in the congress, when all the present delegates went through a real welcome journey visiting political authorities beginning with the mayor of the city and ending at the official house of Juscelino Kubitschek, the Brazilian President. The reporter tells about some relax moments occurred inside the bus during the visits. He mentions that some delegates could communicate themselves using different languages, and sometimes a weird blend of French and English could be heard. He also describes what would have been a passionate dispute between delegates from the United States, France, Cuba, and the Soviet Union about what would be the best alcoholic beverage in the world, each one defending its national product, of course. According to the author, the dispute between the whisky and the vodka was depicted as the hardest one. Once more, we can see that the celebration of an international communion appears side-by-side with a more controversial although subtle narrative that tried to relate the political context of Cold War with the participation of delegates from both sides of the divide world.

10.5 Conclusion

By the example of the eighteenth International Geographical Congress held in Rio de Janeiro in 1956, this chapter tried to explore the politics of location and languages, which is involved in the production of the international rhetoric regarding the so-called international scientific meetings.

The choice of a decentered venue, on the edges of the Anglo-American academic empire, in the words of Minca (2000), should induce changes in the dynamic of congress participation, changes in communication themes, and surely changes in language adoption. If none of these changes occur, the decentered place keeps its status as just another exotic branch company, like a luxurious resort in some distant paradise where you are already familiar with the food, the color of the walls, and with the language you will speak and listen during your pleasant stay. The complexity of the international label that accomplishes good part of congress in geography is nowadays sometimes accompanied by uncomfortable circumstances with different tastes, smells, and sounds.

In a period of an overwhelming hegemony of English communication in geography, it is worthy to talk about an IGU international geographical congress, which offered the impressive number of six different official congress languages. It is more noteworthy considering that it was the last time it happened.

The discussion about language within the scientific discourse is very complex. It is a very hard debate, which largely reaches all of us: on the one hand, those who defend the principle of communication without having regards for anyone and especially the non-natives; on the other hand, those who defend linguistic democracy with a possible risk of disappearing efficient communication. Pretending that this multilingual aspect of the eighteenth IGC was not a controversial feature of the meeting and ignoring that the efficiency of communication can actually be compromised should not necessarily lead us to the path of betting on monolingualism as a way to solve the problem. If there is, thankfully, a substantial and geographically broad critical movement toward "the growing hegemony of English as a global language [that] privileges the geographical discourse of the Anglophone world" (García-Ramon 2003: 1), there is still a lack of more visible or even numerous efforts in order to mitigate the situation regarding the so-called International Geographical Congresses.

We should not only welcome, but also claim to international congress organizers the adoption of short-term measures. This can include an increased use of simultaneous translators during conferences, or the encouragement of the use of abstracts or slides in a second language during communications. In medium-term, in particular, measures to increase the number of bilingual or trilingual proceedings and annals, and in long-term, measures to stimulate learning new languages among the community of monolingual geographers can be included. In addition, the study of some experiences of multilingualism in past can help us in facing the challenge of communication by ourselves in a more symmetric power position and condition.

References

Aalbers M (2004) Creative destruction through the Anglo-American hegemony: a non-Anglo-American view on publications, referees and language. Area 36(3):319–322

Aalbers M, Rossi U (2009) Anglo-American/anglophone hegemony in human geography. In: Kitchin R, Thrift N (eds) International encyclopedia of human geography, vol 1. Elsevier, Amsterdam, pp 116–121

Aranha P (2014) O IBGE e a consolidação da geografia universitária brasileira. Terra Brasilis (Nova Série) 3:2014. https://doi.org/10.4000/terrabrasilis.971

Cameron L, Forsyth I, Yamamura A, Reyes Novaes A (2016) Historical geography as an international discipline 1975–2015: responses. Geogr J 182(3):284–288

Comptes rendus du dix-huitième congrès international de géographie, Rio de Janeiro, 1956. Rio de Janeiro: Comité National du Brésil, 1959–1966, vol 4

Craggs R (2014) Postcolonial geographies, decolonization, and the performance of geopolitics at commonwealth conferences. Singap J Trop Geogr 35(1):39–55

Craggs R, Mahony M (2014) The geographies of the conference: knowledge, performance and protest. Geogr Compass 8(6):414–430

Dalby S (1996) Reading Rio, writing the world: the New York Times and the 'Earth Summit'. Political Geogr 15(6/7):593–613

Desforges L, Jones R (2001) Geographies of languages/languages of geography. Soc Cult Geogr 2 (3):261–264

Dubois JJ (1972) L'UGI et l'organisation des congrès. In: Pinchemel P (ed) Geography through a century of International Congresses. Caen, pp 19–28

García-Ramon MD (2003) Globalization and international geography: the questions of languages and scholarly traditions. Prog Hum Geogr 27(1):1–5

Gutierrez J, Lopez-Nieva P (2001) Are international journals of human geography really international? Prog Hum Geogr 25(1):53–69

Harris CD (1998) Language use in international geographical congresses and the international geographical union. IGU Bulletin 48(2):93–102

Harris CD (2001) English as international language in geography: development and limitations. Geogr Rev 91(4):675–689

Hitchcock C (1957) The eighteenth international geographical congress, Rio de Janeiro, 1956. Geogr Rev 47(1):118–123

Hodder J (2015) Conferencing the international at the world pacifist meeting, 1949. Political Geogr 49:40–50

Hodder J, Legg S, Heffernan M (2015) Introduction: historical geographies of internationalism, 1900–1950. Political Geogr 49:1–6

Kish G (1972) The participants. In: Pinchemel P (ed) Geography through a century of International Congresses. Caen, pp 35–40

Kitchin R (2005) Commentary: disrupting and destabilizing Anglo-American and english-language hegemony in geography. Soc Cult Geogr 6(1):1–15

Legg S (2014) An international anomaly? Sovereignty, the league of nations, and India's princely geographies. J Hist Geogr 43:96–110

Livingstone DN (2007) Science, site and speech: scientific knowledge and the spaces of rhetoric. Hist Hum Sci 20(2):71–98

Livingstone DN (2010) Landscapes of knowledge. In: Meusburger P, Livingstone DN, Jöns H (eds) Geographies of Science. Springer, Dordrecht, pp 3–22
Minca C (2000) Venetian geographical praxis. Environ Plann D Soc Space 18(3):285–289
Pinchemel P (1972) Les congrès et leurs temps. In: Pinchemel P (ed) Geography through a century of international congresses. Caen, pp 185–191
Pumain D (1972) Essai sur les effets de localisation des congrès. In: Pinchemel P (ed) Geography through a century of international congresses. Caen, pp 192–194
Robic M (1996) Résolutions et irrésolutions d'une Cité scientifique; l'entre-deux-guerres. In: Robic M, Briend A, Rössler M (eds) Géographes face au monde. L'Union Géographique Internationale et les Congrès Internationaux de Géographie. Éd. L'Harmattan, Paris, pp 179–226
Robic M (2010) À propos de transferts culturels. Les congrès internationaux de géographie et leurs spatialités. Revue Germanique Internationale 12:33–45
Rössler M (1990) Wissenschaft und Lebensraum. Geographische Ostforschung im Nationalsozialismus. Ein Beitrag zur Disziplingeschichte der Geographie. Reimer, Berlin and Hamburg
Rössler M (1996) 'On political issues' Interview with Chauncy Harris. In: Robic M, Briend A, Rössler M (eds) Géographes face au monde. L'Union Géographique Internationale et les Congrès Internationaux de Géographie. Éd. L'Harmattan, Paris, pp 291–304
Routledge P (2000) 'Our resistance will be as transnational as capital': convergence space and strategy in globalizing resistance. GeoJournal 52(1):25–33
Samers M, Sidaway JD (2000) Exclusions, inclusions, and occlusions in 'Anglo-American geography': reflections on Minca's 'Venetian geographical praxis'. Environ Plann D Soc Space 18:663–666
Shapin S (1998) Placing the view from nowhere: historical and sociological problems in the location of science. Trans Inst Br Geogr 23(1):5–12
Shimazu N (2012) Places in diplomacy: guest editorial. Political Geogr 31(6):335–336
Shimazu N (2014) Diplomacy as theatre: staging the Bandung conference of 1955. Mod Asian Stud 48 (1):225–252
Short JR et al (2001) Cultural globalization, global English, and geography journals. Prof Geogr 53(1):1–11
Volle D (1996a) 'La carte des États: vers une couverture du monde?' In: Robic M, Briend A, Rössler M (eds) Géographes face au monde. L'Union Géographique Internationale et les Congrès Internationaux de Géographie. Éd. L'Harmattan, Paris, pp 41–62
Volle, D (1996b): L'Universalité et ses limites. In: Robic M, Briend A, Rössler M (eds) Géographes face au monde. L'Union Géographique Internationale et les Congrès Internationaux de Géographie. Éd. L'Harmattan, Paris, pp 63–82
Whitehand JW, Edmondson PM (1977) Europe and America: the reorientation in geographical communication in the post-war period. Prof Geogr 29(3):278–282
Withers CWJ (2010a) Geographies of science and public understanding? Exploring the reception of the British association for the advancement of science in Britain and Ireland, c. 1845–1939. In: Meusburger P, Livingstone DN, Jöns H (eds) Geographies of science. Springer, Heidelberg, pp 185–198
Withers CWJ (2010b) Geography and science in Britain, 1831–1939: a study of the British Association for the advancement of science. Manchester University Press, Manchester
Withers CWJ, Finnegan D, Higgitt R (2006) Geography's other histories? Geography and science in the British Association for the advancement of science, 1831–c.1933. Trans Inst Br Geogr 31(4):433–451
Withers CWJ, Higgitt R, Finnegan D (2008) Historical geographies of provincial science: themes in the setting and reception of the British Association for the Advancement of Science in Britain and Ireland, 1831–c. 1939. Br J Hist Sci 41(3):385–415

(Re-) Writing the History of IGU? A Report from the Archives

11

Bruno Schelhaas and Stephan M. Pietsch

Abstract

The *International Geographical Union* (IGU), the international organization for geography, was founded in 1922. In its almost 100-year history, and even in the decades before, the involved geographers produced a great number of printed as well as non-printed materials. Besides the self-experienced and individual IGU history, the written (and visual) tradition offers a unique basis to analyze and reconstruct an important part of international geography in the twentieth and twenty-first centuries. After several destinations and sometimes nebulous life phases, the IGU collection has a new home at the *Leibniz Institute for Regional Geography* in Leipzig. The cataloguing and indexing has just begun. In general, we deal with two kinds of media within the IGU Archive: printed and handwritten textual documents on paper and in digital formats as well and audio-visual documents, a few photographs and, especially, video tapes and digital copies containing interviews. For every kind of these sources, a different kind of storing and analysing has to be developed. This chapter gives an insight into the IGU Archive, especially with a focus on the possibilities and limits of (re-)writing the history of international geography.

Résumé

L'Union géographique internationale (UGI), l'organisation internationale de géographie, a été fondée en 1922. Au cours de son histoire presque centenaire, et même dans les décennies précédentes, les géographes concernés ont produit un grand nombre de documents – imprimés et non imprimés. Outre l'histoire de l'UGI, la tradition écrite (et visuelle) offre une base unique pour analyser et reconstruire une partie importante de la géographie internationale au 20ème et 21ème siècle. Après plusieurs destinations et des phases de vie parfois nébuleuses, la Collection UGI a un nouveau domicile à l'Institut de Géographie Régionale à Leipzig. La description archivistique est toujours en cours. En général, nous traitons deux types de supports au sein de Collection UGI: des documents textuels imprimés et manuscrits sur papier et en format numérique également, et des documents audiovisuels, quelques photographies et surtout des bandes vidéo et des copies numériques contenant des interviews. Pour chaque type de ces sources, il faut développer un type différent de stockage et d'analyse. Ce chapitre

B. Schelhaas (✉) · S. M. Pietsch
Leibniz Institute for Regional Geography, Leipzig, Germany
e-mail: B_Schelhaas@leibniz-ifl.de

S. M. Pietsch
e-mail: S_Pietsch@leibniz-ifl.de

B. Schelhaas et al. (eds.), *Decolonising and Internationalising Geography*, Historical Geography and Geosciences, https://doi.org/10.1007/978-3-030-49516-9_11

donne un aperçu des archives de l'UGI, en particulier en mettant l'accent sur les possibilités et les limites de la (ré)écriture de l'histoire de la géographie internationale.

Resumen

La Unión Geográfica Internacional (IGU), la organización internacional para la geografía, fue fundada en 1922. En sus casi 100 años de historia, e incluso en las décadas anteriores, los geógrafos implicados produjeron una gran cantidad de material, impreso y no. Además de la historia individual de IGU, la tradición escrita (y visual) ofrece una base única para analizar y reconstruir una parte importante de la geografía internacional en los siglos XX y XXI. Después de varios destinos y a veces nebulosos periodos, el archivo de la IGU tiene un nuevo hogar en el Instituto Leibniz de Geografía Regional en Leipzig. La catalogación e indexación acaba de comenzar. En general, tratamos con dos tipos de materiales dentro del Archivo IGU: documentos textuales impresos y escritos a mano en papel y también en formatos digitales, y documentos audiovisuales, algunas fotografías y especialmente cintas de video y copias digitales que contienen entrevistas. Para cada tipo de estas fuentes, se debe desarrollar un tipo diferente de almacenamiento y análisis. Este capítulo ofrece una idea del Archivo IGU, especialmente con un enfoque en las posibilidades y límites de la (re)escritura de la historia de la geografía internacional.

Keywords

International Geographical Union · International science organizations · International relations · Archive history · Source material

11.1 Introduction

This contribution aims to introduce the history of the *International Geographical Union* (IGU), the international science organization for our discipline existing since 1922. The focus is on the collection of IGU itself, but the reader should not expect finished research results or fundamentally new details. We are just at the beginning of a new, or better, continued project, but with a strong connection with the archival sources and its description, and the oral history as well. The collection history mirrors to a high degree of the history of international science relations, influenced especially by the Second World War, the new European post-war order and the Cold War, up to very new documents, which date back only some years. The actual situation of IGU Archives means to take up a new challenge of collecting, organizing and presenting unique documents. Besides this, we have now the chance to open gradually the folders for any kind of future historical research, with all achievements of modern archival description and digital presentation.

Fortunately, we have to note that the state of research regarding the IGU history and of the international congresses is varied and satisfying. Besides the comprehensive studies by Philippe Pinchemel (1972) and Marie-Claire Robic et al. (1996), there is an interesting list of publications, especially connected with George Kish, Geoffrey Martin, Marie-Claire Robic and the working group (E.H.GO) in Paris. In addition, we can especially benefit from Geoffrey Martin's idea to establish a *World Directory of Geographical Archives*, initiated at the meetings of the *Commission History of Geographical Thought* at Bundanoon (Australia) 1988 and Hamburg 1989 (Linke 1992; Martin 1989, 1992), but unfortunately up to now it is unfinished.

When we have a closer look into the literature, we can find several articles about single congresses, commissions and national committees (e.g. Roche 2018), or about international projects like the *International World Map* (Heffernan and Pearson 2015) or the *Bibliographie Géographique International* (Robic 1991). Finally, there are many biographical studies on central figures of IGU history, for example, about Charles Arden-Close (e.g. Close 1947; Freeman 1985), Isaiah Bowman (e.g. Martin 1980; Smith 2003), Emmanuel de Martonne (e.g. Bowd 2012;

Enciu and Ursu 2018; Hallair 2007; Joseph and Robic 1987), Carl Troll (e.g. Huppertz 2000; Winiger 2003; Schenck 2017), Dudley Stamp (e.g. Pilfold 2006) and many others.

Unsurprisingly, our commission, the former *Commission History of Geographical Thought* founded in 1968, offered the institutional as well as practical frame for discussing, analyzing and writing IGU history (Buttimer 1998). This is true up to now: the forthcoming commemorations, the one hundred and fiftieth jubilee of the first *International Geographical Congress* in 2021 and the centenary of the IGU one year later were the reasons to focus on the subject once again. In comparison to all projects before we can now use the archival collection of the IGU without any restrictions.

11.2 IGU History

The *International Geographical Congresses* and the *International Geographical Union* from the very beginning provided a forum for international encounters, for science policy, for joined projects, but also for exclusion, colonial and imperialist ideas. In spite of all, IGU is still active, and today we are one of the leading international science organizations.

The first *International Geographical Congress* met in Antwerp in August 1871. The initial idea was the commemoration of the famous Flemish geographers and cartographers Abraham Ortelius and Gerardus Mercator. Charles Ruelens, director of the Royal Library at Brussels, was the driving force to organize the event (Close 1928: 100–102; Martin 1996: 5–6). Already this congress offered the frame for intellectual and practical discussion, for field trips, for resolutions, publications and international projects. Congresses were followed in no particular order in European cities. The first meeting outside Europe was at Washington (1904), and Cairo was the first hosting city in Africa (1925). These first congresses were highly connected with, and organized by the national Geographical Societies, by universities and academies; there was no central organization or office for international geography.

The situation changed after World War I: The *International Research Council* (IRC) was born in 1918 as a global science organization, but without the Central Powers (Austria, Germany, Hungary, Bulgaria and the Ottoman Empire) as the losers of the war (Greenaway 1996; Schroeder-Gudehus 1978). The National Academies of the Principal Allied Powers, especially Belgium, France and the UK, formed the active group of members, with a main purpose to build international organizations for several research disciplines. The neutral nations joined this group a little later. It is still working as the *International Science Council*.

The *International Astronomical Union* and the *International Union of Geodesy and Geophysics* came already into existence both in July 1919, and they gave important orientation to the geographer's community on their way to build a similar organization. It was mainly the French push to establish an *International Geographical Union* (especially Léon Bellot, the director of French military geography, and Alfred Grandidier, secretary of the *Paris Geographical Society*), initially against strong British opposition (Proposal IGU 1920), but finally successful. The second General Assembly of the IRC in Brussels was the place for the foundation of the *International Geographical Union*. On 27 July 1922, the assembled representatives laid the foundation stone for our international institutional frame, the IGU (Robic 1996).

11.3 IGU Archives I: The Early Years

Because the IGU had no permanent office, it seems to be very likely that the administration and the archives as well moved from one secretary general to his successor. In the first IGU statutes, we find an interesting paragraph in number III, 6: "There shall be an Administrative Bureau of the Union, which, under the direction of the General Secretary, shall conduct the correspondence, administer the funds, preserve the documents, and prepare and issue publications sanctioned by the General Assembly". Following this, the office was located in England (Winchester or London, 1922–1928), Florence (1928–1931), Paris (1931–1938) and Louvain (since 1938). IGU presidents and members of the Executive Committee probably retained additional documents. The limited periods of office prevented a systematic construction of a union archive, too. History of IGU Archives is complicated and sometimes nebulous.

There are some indications that up to the Second World War, and probably in the first post-war years, too, the office and the archives as well were housed in Paris and Louvain in Belgium. It was Emmanuel de Martonne and his network, which played a dominant role within the IGU since the 1930s. Martonne as a long-time secretary general and president; and at Louvain Paul Michotte (Hegenscheidt 1943; Lefèvre 1941), and after his death, Marguerite Alice Lefèvre (Denis 1986; Polspoel 1964) took over as secretary general.

With regard to IGU Archives, there are several but sometimes contradictory indications in the printed sources. One early information can be found in de Martonne's reflection on IGU history in 1950:

> The cataclysm of the Second World War put a stop to all this progress. The archives of the Union, seized and carried off to Berlin, were returned without disorder or loss, but one could not hope to see renewed within a few months the tradition of the congresses and the commissions created by the Union; the spirit of international collaboration, particularly necessary to geographical research, was in danger of being lost. (de Martonne 1950: 5)

Martonne's daughter Héléne supported the details some years later:

> La deuxième guerre mondiale allait ruiner ce bel édifice et réduire les géographes à l'isolement. Dès la fin des hostilités, le Président avait essayé de se mettre en rapport avec les membres du Comité Exécutif. Le Secrétaire Général: Michotte était mort, les archives de l'Union transportée de Louvain à Berlin avaient pu être récupérées sans trop de dommage et Melle Lefèvre avait pris la relève au Secrétariat. (Leconte 1959: 55)

Another secondary source refers to the destination of the collection at Louvain:

> Le pays fut occupé par l'armée allemande; et bientôt la Gestapo se présenta à Heikant-Rotselaar, où séjournait P. Michotte bien malade et près de la mort, pour réquisitionner les archives et la correspondance de l'Union. Vu l'état de santé du sécrétaire-generale et à défaut de celui-ci, la sécrétaire-adjointe fut amenée à Louvain. On comprend l'émotion du pauvre valétudinaire 'soudainement menacé, sinon dans da personne, peut-être dans celle de son adjointe, de représailles brutales'. Heureusement les choses se passèrent mieux que l'on aurait pu le craindre, du moins en ce qui concerne Mademoiselle Lefèvre, car pour P. Michotte le coup était trop dur. Il ne put s'en relever. (Polspoel 1964: 10)

Greenaway mentions this detail as well:

> The Secretariat remained at Louvain. In August 1940 the archives were removed to Germany, but were returned two years later, disordered and defective, but able to preserve some continuity. (Greenaway 1996: 67)

In contrast, Mechtild Rössler presents a slightly different situation, with a focus on a destination in Paris:

> Die Unterlagen aus dem Büro der Internationalen Geographischen Union in Paris wurden beschlagnahmt und verschleppt. (Rössler 1990: 35) When the German troupes came to Paris the SS captured the records of the IGU, which were stored in the « Institut de Géographie » … (Rössler 1996: 313)

Up to this point we have the following list of possible places for the IGU Archives, but with many question marks: London or Winchester, Florence, Paris, Louvain, Berlin, Louvain again, … The situation of the collection between the 1940s and 1980s is up to now unclear. This is also the case regarding a possible destination of the original collection with the documents between 1922 and 1956. We still do not know if it is lost, as written by Geoffrey Martin: "Much of the early material relating to the International Geographical Congresses and the formation of the International Geographical Union now is lost to record." (Martin 1992: 221) or not, as mentioned by George Kish: "The archives of the Union survived the Second World War" (Kish 1989: 111; 1992: 225).

11.4 IGU Archives II: London 1982–2001

The establishment of an "official" and solid union archives took place not until the 1980s. One detail in the sources is of specific interest. In his secretary's report at the eleventh General Assembly 1964 at London, Hans Boesch reported:

> The *archives*, which were housed in the American Geographical Society in New York, were checked by the Secretary in 1963. It was found that their content was not of a lasting value and covered only the period of 1949–1956 (Prof. Kimble): In conference with Prof. Kimble it was decided to dispose of the assembled material; a few documents were taken to Zurich. (IGU Bulletin 16(1965)1: 14–15)

In conclusion, we can follow that IGU Archives (or parts of it) was immediately connected with the office of the General Secretary and Treasurer. This seems to be true for George Kimble (1949–1956) with his various academic stations, but closely connected with the American Geographical Society, where he found a place for the collection. Hans Boesch and Chauncy D. Harris probably followed this example.

Finally, the *Royal Geographical Society* in London was elected as IGU Archives in 1982. It was the beginning of the professional management, of course with minor and major problems, but formally written down in the IGU statutes. We can find the first note in the IGU Bulletin in 1981:

> Arrangements are now in progress to establish a permanent base for the IGU Archives at the Royal Geographical Society, 1 Kensington Gore, London SW7. The assembly of the records should be complete by the end of 1981: the material deposited will then be available for examination by bona fide students of the history of geography. Editors of the publications of IGU Commissions and Working Groups are asked to ensure that a record copy is sent to the IGU Archives at the Royal Geographical Society. (IGU Bulletin 32, 1981, 2: 4)

The driving force to establish IGU Archives was Past President Michael J. Wise (Johnston and Board 2018). The agreed rules of contribution worked only occasionally, but the archival documents had a save home for more than one decade. In 1983, the IGU officially announced:

> The IGU Archive has now been established at the Royal Geographical Society [...] and is open to use for research purposes by scientists and scholars. Those intending to use the archive should previously notify the Deputy Director of the Royal Geographical Society so that appropriate arrangements can be made.
> Chairmen of Commissions and Working Groups are requested to send copies of all publications, including maps, to the Archive. Important unpublished papers which throw light on the development of research work and of international collaboration in the study of the subject should also be sent. (IGU Bulletin 33, 1983, 1/2: 6)

In several following IGU documents, we can find information and appeals concerning IGU Archives, for example, in the *Guidelines for chairpersons of commissions and study groups*:

> All chairpersons are advised to either transfer their files at the end of each term to the R.G.S. or at least send them a letter indicating where the files have been deposited. The latter applies if a commission decides to deposit old files with the university or a local archive. However, we would like to have a central register of archival material for future historians of science to be able to draw on. The same applies to publications issued by each commission and study group. (IGU Guidelines for chairpersons of commissions and study groups 1989, 1993, 1994)

Other references can be found in the *IGU Bulletin*:

> Although the IGU Secretariat has established an archive which is deposited with the Royal Geographical Society in London, mostly central files for 1956-88 are kept there. Disposal of other files was never standardized and each chairperson made his/her own decision. It is expected that in due

> course a comprehensive index of geographical archives will be established. (IGU Bulletin 40, 1990, 1/2: 7)
> The archives of the IGU are deposited with the Royal Geographical Society in London. They contain mostly central files for 1956-88. Disposal of other files was never standardized and each chairperson made his/her own decision. It is expected that in due course a comprehensive index of geographical archives will be established. As well, the IGU participates in an ICSU-sponsored collection of documents kept in Paris. (IGU Bulletin 42, 1992, 1/2: 8)

The agreed rules of contribution worked only occasionally, but the archival documents from 1956 up to the 1980s had a safe and famous home for more than one decade. The compliance of the rules to hand over relevant documents is still a decisive problem up to the present days; it is a permanent problem of archival collections of any kind.

11.5 IGU Archives III: Rome, 2001–2013

Around the turn of the millennium, a second phase of IGU Archives began. The *Home of Geography* was the attempt to establish a central IGU office, including a library and IGU Archives. The collection moved in 2001 from London to Rome and found a new home in the *Villa Celimontana*, the main office of the *Italian Geographical Society*. Already in November 2000, secretary general Eckart Ehlers delivered 29 boxes of additional documents concerning his time of office from Bonn to Rome.

The objectives of the *Home of Geography* were very ambitious, and not all points could be realized. However, IGU Archives was again in the focus, and an important decision to develop the collection was made: It was Geoffrey Martin who organized and catalogued the archives in 2002, and who worked out the basics for the actual new cataloguing. He archived 214 boxes and developed a useful structure of the collection with twelve series:

I. The meaning of the IGU
II. Member countries of the IGU and correspondence relating thereto
III. Congresses of the IGU
IV. Regional Conferences of the IGU
V. IGU Commissions
VI. IGU Study Groups
VII. IGU Working Group
VIII. IGU Executive
IX. IGU Newsletter, Bulletin and other publications
X. Affiliation with other organizations
XI. Financial income and expenditure of the Union
XII. Miscellaneous correspondence.

In the following decade, several boxes with new archival material came to Rome, including documents from commissions, congresses and from *Home of Geography* itself; in 2007, the RGS delivered 12 boxes containing the Thomas Walter Freeman papers, mainly concerning our *Commission History of Geographical Thought* and the *Biobibliographical Studies*. In spite of all progress, almost everything of the new accessions remained in the boxes and was not arranged according to Geoffrey Martin's structure.

11.6 IGU Archives IV: Leipzig, 2013

In 2011, the situation changed again: the *Home of Geography* was closed and the cooperation with the *Italian Geographical Society* could not be continued. IGU Archives needed a new home again. The IGU Executive Committee decided to transfer the collection to the *Leibniz Institute for Regional Geography* at Leipzig, and the boxes arrived in May 2013 at our institute. IGU Archives is now at Leipzig surrounded by many documents concerning the history of modern geography. Actually, we are busy in organizing and cataloguing the archives: an expensive task and heavy responsibility. However, we are hopeful to manage this task before the IGU jubilee in 2021/22.

The collection contains a large number of varied documents, different kinds of media and with various importance for reconstructing history of international geography. In a second part, some typical examples will introduce the character of IGU Archives and its opportunities for future historical research.

11.7 Spotlights from IGU Archives

Based on the overview about the history of IGU, its archives and the preliminary structure of the collection elaborated by Geoffrey Martin in 2002, the following part will introduce the variety of material of IGU Archives. These insights, in turn, represent some of the results of the works undertaken at the *Leibniz Institute for Regional Geography* at Leipzig in 2016 and 2017 in the context of the IGU Jubilee Task Force.

Before getting started with the introduction of the material, some initial thoughts regarding practical questions of structuring and (re-)ordering the IGU Archives seem to be crucial. As already mentioned, the documents are pre-structured in twelve series, stored in 214 boxes. Nevertheless, to open a box often means to be confronted with a kind of surprise bag. This means, that especially in the case of so-called uncategorized boxes, the variety of different content can be extremely vast. This evidence shows the need for a more differing archival system. Therefore, the actual archival work at the *Archive for Geography* in Leipzig is especially based on two principles: firstly, to develop a user-friendly finding aid and to offer convenient using conditions, and secondly, to regard the historical authenticity of the documents and to follow the so-called principle of provenance as much as possible (Brenner-Wilczek et al. 2006: 35). Because IGU Archives is now based in Germany, any use of the documents depends on German laws of data protection and archival work, which means that some "younger" sources—especially digital documents or very new correspondence are probably not fully accessible.

Overall, four main categories of material can be distinguished within the collection until now. The first one is that of textual documents, which are hand- or typewritten. This category mainly consists of diverse correspondences—also some post- and greeting cards, which are of some more personal content than the official documents—invoices, minutes and communiques as well as already printed material. We can find for instance IGU Newsletters and Bulletins, but also whole book chapters or articles.

The second category is that of visual data. Besides different photographs—portraits of popular geographers as well as images of landscapes—also up to now uncounted number of maps are included here.

Another category of sources is that of audio-visual material—a unique pool of sources for any analysis of international history of geography of the last five decades. Especially, the original video tapes of the *Invitation to Dialogue* Project, which was established by Anne Buttimer and Torsten Hägerstrand in cooperation with the *University of Lund* in the 1970s and 1980s (International Geographical Union 2014b), can be found in IGU Archives. The videos provide interesting insights on the global development of modern geography from multiple perspectives, as they consist of interviews with prominent contemporary geographers from all around the world like, for example, David Harvey, Akin Mabogunje, María Teresa Gutiérrez de MacGregor, Patrick Armstrong, Keiichi Takeuchi, etc. In addition to the physical storage of the recordings in Leipzig, the videos are already available on IGU's *YouTube* channel (International Geographical Union 2014c).

The last category is the digital documents, mainly several text files, e.g. invoices, financial balances and certainly email communication, saved on CD ROM, DVD, hard or flash drive. In the following chapter, examples of two categories, textual documents and visual data, illustrate the varied kind of sources and their importance for historical research.

11.8 Textual Documents

As far as we know at this point, hand- or typewritten documents cover most of the physical material within IGU Archives. On the one hand, correspondence about different topics—IGU related, private and mixed—can be found in all folders. Most of these documents deal directly with IGU issues, e.g. the organization of *International Geographical Congresses*; the design of subject-related commissions (for an overview see International Geographical Union 2014a) or the processing of different publications. Besides this, many of these documents show interesting insights into the social life of an international science organization. On the other hand, there is a huge amount of finance-related material, especially in the folders of the National Committees from 1956 to 1992, with invoices, receipts and administrative forms. This group of sources offers unique possibilities for reconstructing the structural environment of the institution, including the different methods of administration and management under changing political and scientific conditions.

The following examples illustrate the specific character of the documents within IGU Archives. Figure 11.1 shows a letter from IGU President Shiba Prasad Chatterjee to General Secretary Hans Boesch from 1966 about planned journeys. Besides the dates and structural facts, it is easily possible to learn something about the social history of IGU, about the relationship between president and his secretary and the social character of their relation. The following quote from Chatterjee's letter underlines the multiple meaning of this single written document: "It goes without saying that we remain firm in our friendship, even though we may differ sometime in certain things".

The second example (Fig. 11.2) is equally characteristic for this group of sources. It is an invoice for the annual subscription of Brazil to the IGU for the year 1960 with the amount of 1.500 US$ and its equivalent 6.450 SFrs.

Besides the folders related to the National Committees, the IGU Commissions form another important chapter within IGU Archives, with a lot of information about the practical work within IGU. Especially, the *Commission on the History of Geographical Thought* is documented in several boxes, with most interesting documents concerning history of geography and its professional elaboration. As one example, there is a huge amount of documents, regarding the publication series of the commission, *Geographers: Biobibliographical Studies*. In a hitherto unsorted box from 1984, we can find different types of texts, which were produced and communicated during the editing process of volume 8. Firstly, the box contains correspondence of Thomas Walter Freeman, the editor of the series, to Samuel John Kenneth Baker about the idea of writing a contribution about John Harald Wellington. Secondly, it contains a variety of sources, Baker used for his article, e.g. index cards or several copies of publications of Wellington. Thirdly, also the final article can be found, so that a reconstruction of the whole process of production of an article from the first idea to its publication can be undertaken. This example reveals important information on the publication strategy of the IGU in the 1980s as well.

11.9 Visual Data

Besides the textual documents, visual data is another important category of archival material —even if not in a quantity as the written content —within IGU Archives. In the first place, this kind of media can be perfectly used for illustrating future works on IGU history, in particular for elaborating more convenient and readable works for the scientific community, and a broader audience, too. But the value of visual material goes far beyond this illustrational aspect. As Heike Talkenberger already mentioned in 1994, we are living in a culture, which is largely dominated by visuality; also historical awareness is predominantly shaped by visual elements (Talkenberger 1994: 289). Moreover, especially photographs can provide considerable more information, as they possibly provide insights, which cannot be extracted from classical texts (Paul 2006: 9). Therefore, it totally makes sense to analyze these pictures as independent sources, for example, with regard to the social status of key persons of an international organization, which possibly could be interpreted out of the respective style of clothing.

With regard to the *International Geographical Congresses*, a unique series of photographs from

E 31.1.66

UNIVERSITY OF CALCUTTA

Phone : 47-5144

Prof. Shiba P. Chatterjee
Head of Geography Department
President,
International Geographical Union

35, Ballygunge Circular Road,
Calcutta-19 (India)

Dated January 25, 1964

My dear Boesch,

So pleased to receive your letter of January 20, which you posted from the Santa Cruz air port. I was happy to receive you in Calcutta and wished that you could extend your sojourn in India and in fact I arranged it accordingly with the help of the U.G.C. But I quite realised why you could not stay longer. It goes without saying that we remain firm in our friendship, even though we may differ sometime in certain things. To be away from family is something one does not like, and in addition to that if the country is far away from home one feels still more miserable. I am glad that you enjoyed your trip to India despite these two facts. It shows that you are a 'non-conformist' per excellence. As I told you, I am planning to go to Moscow on March 29 and then to visit a few countries before reaching Paris on April 24, where I am expected to deliver lectures at the University besides attending the Commission on National Atlases. On my return trip to India I am planning to go to Madrid, London, Dublin, Bruxelles, Australia and Warsaw, and then again to Moscow before returning to India sometime by the end of June 1966 . I hope you will make it convenient to come to Paris when I am there.

Mr. Bassols of Mexico is staying with me now. He tells me that it may be worthwhile to visit a few Latin-American countries when I go to Mexico in August. A visit to Habana, Guatimala Tegucigalpa, Panama, Bogota and Caracas is suggested. I shall let you know more about the trip in my next letter. Please keep me informed of your plan of work.

I believe you have now settled down to your multifarious activities in Zurich and my sincere advice to you is that you should not strain yourself too much.

With my best regards,

Yours sincerely,

S.P. Chatterjee

S P Chatterjee

Prof. Hans Boesch
Geographisches
Institut der Universitat
Blumlisalpstrasse 10
8006 Zurich
Switzerland.

P.S. We just received a very bad news • Air India Plane crashed into Mt. Blanc range. It is very sad that Dr. Bhabha was in the plane and was among the victims.

Fig. 11.1 Letter of Chatterjee to Boesch, 25th January 1966. *Source* IGU Archives VIII-1-2

Union Géographique Internationale — International Geographical Union

Secrétaire-Trésorier — Secretary-Treasurer:
Prof. Hans Boesch
Geographisches Institut der Universität
30, Freiestrasse
Zurich 32 — Suisse/Switzerland

Professor Hilgard O'Reilly Sternberg
Centro de Pesquisas de Geografia
do Brazil,
Av. Presidente Antônio Carlos 40- 9°
Rio de Janeiro

FACTURE – INVOICE

Zurich, January 19, 1960

Pays:
Country: B r a z i l

Catégorie:
Category: VIII

I. A l'UGI, contribution pour l'année financière:
To IGU, contribution for the financial year: 1960
(1XXXXXXXXXXXXXXX)

II. Divers:
Various:

	US $	ou/or	SFrs.
I.	1'500		6'450.-
Total:	US$ 1'500	ou/or	SFrs. 6'450.-

Le payement peut être effectué soit par chèque à l'ordre de l'Union Géographique Internationale, Zurich, en l'envoyant à l'adresse mentionnée ci-dessus, soit par versement direct pour le compte de l'Union au Crédit Suisse, Zurich.

Payment can be made either by cheque to the order of International Geographical Union, Zurich, and sent to the above address, or by remittance to the account of the Union with the Swiss Credit Bank, Zurich.

Fig. 11.2 Invoice for National Committee of Brazil, 19th January 1960. *Source* IGU Archives II-1-3

Fig. 11.3 Congress participants during a chat. *Source* IGU Archives III-2-2-IGC Montreal

the conference in Montreal in 1972 underlines the opportunities of visual data for reconstructing history of geography. Certainly, official portraits can be used to illustrate future publications about IGU history, in general or directed to the Montreal Congress in 1972. Another group of photographs shows events and moves during the congress. As depicted in Figs. 11.3 and 11.4, these snapshots are not only recording the official happenings. In contrast, they show everyday-life practices of scientific conferences, e.g. a chat between participants, possibly in-between exhausting panel sessions, or even participants waiting for already chartered buses, which would take them to interesting locations around Montreal in the context of geographical excursions. Going beyond this interpretation of historical images, even some approaches in terms of a *visual history* seem to be possible. In this regard, critical questions, which are focusing on the deeper meanings of the photographs, including the levels of representation and reception (Paul 2006: 27), could be raised.

11.10 Outlook—Towards a Source-Based History of the IGU

Certainly, these selected examples from IGU Archives direct to a very small part of IGU history only, but they illustrate the great variety of material and its meaning for historical research. Therefore, it seems to be obvious that the first step within writing of a new IGU history has to be the professional new archival description of IGU Archives, including ordering, cassations and indexing of the material. It is also useful to consider to undertake a partly digitalization of the material in order to secure the, in some cases, almost damaged sources, and to offer an uncomplicated access to all possible users. In spite of some limitations of using, the collection offers diverse possibilities for writing a source-based IGU history.

Besides the limitations in terms of data protection or damaged documents, we have again to note the gaps within the collection and the loss of

Fig. 11.4 Excursion buses. *Source* IGU Archives III-2-2-IGC Montreal

the old material before 1956. Further investigations of several collections in different countries are necessary, as well as an incorporation of oral IGU history is needed. Writing a new source-based IGU history should include a critical evaluation of the sources, and focus on the so-called intrinsic value of the material. When overlooking the material, it is easy to discover that the majority of documents consists of formal files shaped in a western kind of administrative and scientific thinking. They document ways of international cooperation, integration and exclusion, with several scientific as well as political or colonial motivations, but mostly resulting from bureaucratic procedures, especially concerning formal and financial topics. The composition of documents with many administrative recordings, correspondence, congress material and audio-visual documents offers a basis to new insights into IGU history and history of international geography in general. This is especially true with regard to the social history of international geography, and referring to the visual data for a new approach of a kind of a *visual history* of geography. Photographs can also be used as helpful triggers within interviews with IGU witnesses, e.g. concerning *International Geographical Congresses* or other IGU events.

The present kind of archival material opens another possible topic, the quantitative versus qualitative approach of writing IGU history. In a quantitative sense, the files provide a multitude of numerical data, not only about the financial situation of the organization, member countries, commissions and cooperation with other institutions. This group of documents can be used to elaborate a number of interesting maps, which visualize the global interconnections of the member countries of the IGU in the twentieth century. In a qualitative sense, personal relations between individuals, for example, presidents of the IGU and their secretaries can be extracted from the correspondence. Some of these letters can also be used to analyse the international political situation during the Cold War, and to study the reaction of an international scientific organization on this political change. Many written documents from the 1990s can be found in the collection, dealing with the reorientation after the fall of the Iron Curtain. Again, the documents offer very interesting sources for reconstructing the younger history of our subject.

References

Bowd G (2012) Un géographe français et la Roumanie. Emmanuel de Martonne, 1873–1955. Harmattan, Paris

Brenner-Wilczek S, Cepl-Kaufmann G, Plassmann M (2006) Einführung in die moderne Archivarbeit. Wissenschaftliche Buchgesellschaft, Darmstadt

Buttimer A (1998) Geography's contested stories. Changing states-of-the art. Tijdschrift voor Economische en Sociale Geografie 89(1):90–99

Close C (1928) Address at the anniversary general meeting. Geogr J 72(2):97–116

Close C (1947) Geographical by-ways and some other geographical essays. Arnold, London

Denis J (1986) Marguerite Alice Lefèvre 1894–1967. Geogr Biobibliograph Stud 10:106–110

Enciu N, Ursu VG (2018) Emmanuel de Martonne – l'avocat de la Roumanie à la Conférence de Paix à Paris en 1919–1920 en matière de Bessarabie. Éditions universitaires européennes, Saarbrücken

Freeman TW (1985) Charles Frederick Arden-Close. Geogr Biobibliograph Stud 9:1–14

Greenaway F (1996) Science international. History of the International Council of Scientific Unions. Cambridge University Press, Cambridge

Hallair G (2007) Le géographe Emmanuel de Martonne et l'Europe centrale. CNRS-Prodig, Paris

Heffernan M, Pearson AW (2015) Globalizing cartography? The international map of the world, the International Geographical Union, and the United Nations. Imago Mundi 67(1):58–80

Hegenscheidt A (1943) Le géographe Paul Michotte. UCL, Louvain

Huppertz J (2000) Carl Troll zum 100. Geburtstag, Siebenberg, Waldeck

International Geographical Union (2014a) Commissions. Online: https://igu-online.org/organization/commissions/. Access 1 Oct 2018

International Geographical Union (2014b) Dialogue project. Online: https://igu-online.org/dialogue-project/. Access 24 Sept 2018

International Geographical Union (2014c) IGU channel. Online: https://www.youtube.com/channel/UC1WzSi02jYP3QgjseHxKB3g/feed. Access 24 Oct 2018

Johnston R, Board C (2018) Michael John Wise (1918–2015). Geogr Biobibliograph Stud 36:51–88

Joseph B, Robic MC (1987) Exploration d'archives: autour des papiers d'Emmanuel de Martonne. Acta Geogr 72:37–65

Kish G (1989) International Geographical Union. A brief history. IGU Bull 39(2):110–119

Kish G (1992) International Geographical Union. A brief history. GeoJournal 26(2):224–228

Leconte H (née de Martonne) (1959) Histoire de l'Union Géographique Internationale et des Congrès Internationaux de Géographie. IGU Newsl 10(1):3–20 and (2):43–69

Lefèvre MA (1941) In memoriam Paul Lambert Michotte 1876–1940. L. Durbeck, Louvain

Linke M (1992) Memorandum. The idea of a "world directory of geographical archives". GeoJournal 26 (2):222

Martin G (1980) The life and thought of Isaiah Bowman. Archon Books, Hamden, Conn

Martin G (1989) Toward "a world directory of geographical archives". IGU Bull 39(2):120–124

Martin G (1992) History of geography archives—a world directory. GeoJournal 26(2):221–222

Martin G (1996) 125 years of geographical congresses and the formation of the International Geographical Union. Or, from Antwerp to the Hague. IGU Bull 46:5–26

de Martonne E (1950) Brief history of the International Geographical Union. IGU Newsl 1(2):3–5

Paul G (2006) Von der Historischen Bildkunde zur Visual History. Eine Einführung. In: Paul, G (ed) Visual History. Ein Studienbuch. Vandenhoek & Ruprecht, Göttingen, pp 7–36

Pilfold WE (2006) Sir Laurence Dudley Stamp (1898–1966), geographer and public servant. A critical biography. Ph.D. Thesis University of Sussex, Brighton

Pinchemel P (ed) (1972): Geography through a century of international congresses. Caen

Polspoel GL (1964) La carrière et l'activité scientifique de Mademoiselle M. A. Lefèvre. Professeur à l'Université de Louvain. Acta geographica lovaniensa 3:5–21

A Proposal for an International Geographical Union (1920). Geograph J 55(2):139–142

Robic MC (1991) La Bibliographie géographique (1891–1991), témoin d'un siècle de géographie: quelques enseignements d'analyses formelles. Annales de géographie 561(2):521–577

Robic MC (1996) La naissance de l'Union Géographique Internationale. In: Robic MC, Briend AM, Rössler M (eds) Géographes face au monde. L'Union Géographique Internationale et les Congrès Internationaux de Géographie. Éd. L'Harmattan, Paris, pp. 23–39

Robic MC, Briend AM, Rössler M (eds) (1996) Géographes face au monde. L'Union Géographique Internationale et les Congrès Internationaux de Géographie. Éd. L'Harmattan, Paris

Roche M (2018) From an "Absent to a silent presence". New Zealand at the international geographical congress 1875 to 1952. NZ Geogr 74(2):1–9

Rössler M (1990) "Wissenschaft und Lebensraum". Geographische Ostforschung im Nationalsozialismus. Ein Beitrag zur Disziplingeschichte der Geographie. Reimer, Berlin and Hamburg

Rössler M (1996) Archives de et sur l'UGI. In: Robic MC, Briend AM, Rössler M (eds) Géographes face au monde. L'Union Géographique Internationale et les Congrès Internationaux de Géographie. Éd. L'Harmattan, Paris, pp 313–316

Schenck W (ed) (2017) Carl Troll. Dokumente zu seine Biographie und seinem wissenschaftlichen Werk. Ferger, Bergisch Gladbach

Schroeder-Gudehus B (1978) Les scientifiques et la paix. La communauté scientifique internationale au cours des années vingt. Pr. de l'Univ. de Montréal, Montréal

Smith N (2003) American Empire. Roosevelt's geographer and the prelude to globalization. University of California Press, Berkeley

Talkenberger H (1994) Von der Illustration zur Interpretation. Das Bild als historische Quelle. Methodische Überlegungen zur Historischen Bildkunde. Zeitschrift für Historische Forschung 21(4):289–313

Winiger M (ed) (2003) Carl Troll. Zeitumstände und Forschungsperspektiven. Asgard, Sankt Augustin

The manufacturer's authorised representative in the EU is Springer Nature Customer Service Centre GmbH, Europaplatz 3, 69115 Heidelberg, Germany. If you have any concerns regarding our products, please contact ProductSafety@springernature.com

Printed and bound by CPI Group (UK) Ltd, Croydon, CR0 4YY
28/09/2026
02224078-0011